AXING THE TAX

AXING THE TAX

The Rise and Fall of Canada's Carbon Tax

FRANCO TERRAZZANO

Sutherland House
416 Moore Ave., Suite 304
Toronto, ON M4G 1C9

First edition, April 2025

If you are interested in inviting one of our authors to a live event or media appearance, please contact sranasinghe@sutherlandhousebooks.com and visit our website at sutherlandhousebooks.com for more information.

We acknowledge the support of the Government of Canada.

Manufactured in Canada
Cover designed by Leah Ciani and Jordan Lunn

Library and Archives Canada Cataloguing in Publication
Title: Axing the tax : the rise and fall of Canada's carbon tax / Franco Terrazzano.
Names: Terrazzano, Franco, author.
Description: Includes bibliographical references and index.
Identifiers: Canadiana (print) 20250158329 | Canadiana (ebook) 2025015837X | ISBN 9781998365654 (softcover) | ISBN 9781998365661 (EPUB)
Subjects: LCSH: Carbon taxes—Canada. | LCSH: Carbon taxes—Political aspects—Canada. | LCSH: Taxation—Canada.
Classification: LCC HJ2449 .T47 2025 | DDC 336.200971—dc23

ISBN 978-1-998365-65-4
eBook 978-1-998365-66-1

For my Dad. Thank you for all you did.

CONTENTS

CONTENTS

PROLOGUE

Every now and then, the underdog wins one.

And it looks like that's happening in the fight against the carbon tax.

It's not over yet, but support for the carbon tax is crumbling. Some politicians vow to scrap it. Others hide behind vague plans to repackage it. But virtually everyone recognizes support for the current carbon tax has collapsed.

It wasn't always this way.

For about a decade now, powerful politicians, government bureaucrats, academics, media elites and even big business have been pushing carbon taxes on the people.

But most of the time, politicians never asked the people if they supported carbon taxes. In other words, carbon taxes, and the resulting higher gas prices and heating bills, were forced on us.

We were told it was good for us. We were told carbon taxes were inevitable. We were told politicians couldn't win elections without carbon taxes, even though the politicians that imposed them didn't openly run on them. We were told that we needed to pay carbon taxes if we wanted to leave a healthy environment for our kids and grandkids. We were told we needed to pay carbon taxes if we wanted to be respected in the international community.

In this decade-long fight, it would have been understandable if the people had given up and given in to these claims. It would have been easier to accept what the elites wanted and just pay the damn bill. But against all odds, ordinary Canadians didn't give up.

Canadians knew you could care about the environment and oppose carbon taxes. Canadians saw what they were paying at the gas station and on their heating bills, and they knew they were worse off, regardless of how many politicians, bureaucrats, journalists and academics tried to convince them otherwise. Canadians didn't need advanced degrees in economics, climate science or politics to understand they were being sold a false bill of goods.

Making it more expensive for a mom in Port Hope to get to work, or grandparents in Toronto to pay their heating bill, or a student in Coquitlam to afford food won't reduce emissions in China, Russia, India or the United States. It just leaves these Canadians, and many like them, with less money to afford everything else.

Ordinary Canadians understood carbon taxes amount to little more than a way for governments to take more money from us and dictate how we should live our lives. Ordinary Canadians also saw through the unfairness of the carbon tax.

Many of the elites pushing the carbon tax—the media, politicians, taxpayer-funded professors, laptop activists and corporate lobbyists—were well off and wouldn't feel the brunt of carbon taxes. After all, living in a downtown condo and clamouring for higher carbon taxes doesn't require much gas, diesel or propane.

But running a business, working in a shop, getting kids to soccer and growing food on the farm does. These are the Canadians the political class forgot about when pushing carbon taxes. These are the Canadians who never gave up. These are the Canadians who took time out of their busy lives to sign petitions, organize and attend rallies, share posts on social media, email politicians and hand out bumper stickers.

Because of these Canadians, the carbon tax could soon be swept onto the ash heap of history.

I wrote this book for two reasons.

The first is because these ordinary Canadians deserve it. They worked really hard for a really long time against the odds. When all the power brokers in government told them, "Do what we say—or pay," they didn't give up. They deserve to know the time and effort they spent fighting the carbon tax mattered. They deserve all the credit.

Thank you for everything you did.

The second reason I wrote this book is so people know the real story of the carbon tax. The carbon tax was bad from the start and we fought it from the start. By reading this book, you will get the real story about the carbon tax, a story you won't find anywhere else.

This book is important because if the federal Liberals' carbon tax is killed, the carbon taxers will try to lay blame for their defeat on Prime Minister Justin Trudeau. They will try to say that carbon taxes are a good idea, but Trudeau bungled the policy or wasn't a good enough salesman. They will try to revive the carbon tax and once again make you pay more for gas, groceries, and home heating.

Just like with any failed five-year plan, there is a lingering whiff among the laptop class and the taxpayer-funded desk rulers that this was all a communication problem, that the ideal carbon tax hasn't been tried yet. I can smell it outside my office building in Ottawa, where I write these words. We can't let those embers smoulder and start a fire again.

This book shows why the carbon tax is, and always will be, bad policy for ordinary Canadians.

Franco Terrazzano
Federal director, Canadian Taxpayers Federation

CHAPTER 1

THE ROOTS OF THE CARBON TAX

Let's look back at the history of the carbon tax and start with an interesting piece of political trivia: Justin Trudeau wasn't the first prime minister to push a carbon tax—it was Joe Clark.

Clark's Progressive Conservative government floated a similar idea in the 1979 federal budget. Clark tabled a plan to increase the gasoline tax by eighteen cents per gallon "to promote conservation and raise revenues." In its budget speech, the minority Clark government said this tax hike was needed due to "the absolute necessity of further encouraging our people to use fewer oil products."

But the proposal did not go over well.

"It is the budget that stole Christmas," said an opposition Liberal spokesman.

Social Credit leader Fabien Roy said he could "not vote for a budget that provided for an increase in gasoline taxes."

As a result, Clark's budget, and his oil use reduction tax, did not come to pass. His government quickly lost the confidence of the House of Commons and was brought down.

The New Democrats went so far as to attack the Clark government "for its outright betrayal of its election promises to lower interest rates, cut taxes and to stimulate the growth of the Canadian economy."

Soon after, former prime minister Pierre Trudeau came out of retirement and defeated Clark in the 1980 election. Clark lost thirty-three seats in the House of Commons and the Liberals formed a majority government.

Fast-forward a few decades and proponents of carbon taxes reappeared among Canada's green activists and academics. The David Suzuki Foundation claims to have been calling for carbon taxes since 1998. Canada's Ecofiscal Commission, a collection of academics, has been writing reports promoting carbon taxes since 2015.

Even the Conservative Party of Canada has not been immune to the influence of carbon tax advocates.

In 2004, Stephen Harper's Conservatives promised to "investigate a cap-and-trade system that will allow firms to generate credits by reducing smog-causing pollutants."

In its 2008 election platform, the Conservatives pledged to "develop and implement a North America-wide cap-and-trade system for greenhouse gases and air pollution."

A cap-and-trade scheme is another form of carbon tax. The government sets up an artificial market and mandates a certain level of emissions that decline over time. The government-mandated emissions cap determines the carbon tax a company must pay to buy credits. Under cap and trade, companies pass the costs onto consumers through higher prices at gas stations and on home heating bills.

As the Carbon Tax Center notes: "Politically, cap-and-trade has functioned as a 'safe harbour' for politicians who grasp the need to price carbon emissions but cling to the need to 'hide the price' to appease interest groups and/or voters."

The idea of a carbon tax is much older than even Clark's proposal of 1979.

In fact, the roots of carbon taxes trace all the way back to a paper written in 1920 by the economist Arthur Pigou. In the paper, entitled the *Economics of Welfare*, Pigou argued that a tax on things like air pollution is necessary because of what's known as "externalities."

An externality occurs when a transaction is conducted between two participants, usually a business and a customer, which causes a third party to be impacted. For example, a positive externality could be smelling a freshly baked pie cooling on a windowsill as you walk by a house. You didn't pay for the pie, but you still get to enjoy the pleasant smell. On the other hand, pollution is an example of a negative externality.

CHAPTER 2

FEDERAL FLIRTATIONS WITH CARBON TAXES

To his credit, former Liberal leader Stéphane Dion was a rare exception among politicians because he was bold enough to explicitly run on a carbon tax during an election.

But voters soundly rejected his proposal.

Dion's Green Shift was his cornerstone campaign promise during the 2008 federal election.

But the Liberals were far from the only major political party in Canada to flirt with implementing a carbon tax.

As mentioned above, the Conservative Party's 2004 election platform said it would investigate a cap-and-trade carbon tax. Four years later, its 2008 election platform went further, promising a Conservative government would "develop and implement a North America-wide cap-and-trade system for greenhouse gases and air pollution."

The inclusion of "North America-wide" is key to understanding this promise. The Conservatives had included this policy proposal in large part because President Barack Obama was talking about implementing a cap-and-trade system. In fact, both major presidential candidates in the 2008 US election came out in favour of cap-and-trade carbon taxes. But by 2010, Obama had been unable to get a cap-and-trade system passed. As a result, in 2011, Prime Minister Stephen Harper confirmed the Canadian government would not implement a cap-and-trade carbon tax.

The New Democrats had also included a cap-and-trade carbon tax in both its 2008 and 2011 campaign platforms.

But only the Liberals were planning to go all-in by implementing a consumer carbon tax. Dion's 2008 Green Shift would have hiked his carbon tax to cost $40 per tonne of greenhouse gas emissions over four years. (For context, as of April 1, 2024, the Trudeau carbon tax costs $80 per tonne and is scheduled to keep rising until 2030.)

Dion's experience pushing his Green Shift on Canadians should have served as a warning to Prime Minister Justin Trudeau. Harper made hay of the proposal in the 2008 election by comparing it to Pierre Trudeau's National Energy Program and the disastrous effect it had on Western Canada.

"This is different in that this will actually screw everybody across the country," Harper said. "[A carbon tax] will recklessly harm the economy and the economic position of every Canadian family."

Harper wasn't the only one comparing Dion's carbon tax to the National Energy Program.

"I, for one, wrote two articles trying to warn them that if they did not change the structure of their carbon tax to address its unfairness to Western Canada, they risked rekindling the western alienation and regional discontent that followed the National Energy Program," wrote Janice Mackinnon, Saskatchewan's former New Democratic Party (NDP) finance minister.

MacKinnon added that Dion's carbon tax also faced "opposition to it from several premiers, including some Liberal ones."

Back then, the NDP knew a consumer carbon tax would make life more expensive for ordinary Canadians.

"Those advocating a carbon tax suggest that by making the cost of certain things more expensive people will make different choices, but Canada is a cold place and heating your home really isn't a choice," said federal NDP leader Jack Layton. "We shouldn't punish people and that's what a carbon tax does."

This mirrored the NDP's position in the provinces. When the British Columbia Liberals created the first carbon tax in Canada in 2008, the NDP railed against it. Its leader at the time, Carol James, called claims of a revenue-neutral carbon tax "lipstick on a pig." The BC NDP's 2008 campaign slogan was "Axe the Tax."

The Canadian Taxpayers Federation (CTF) launched a website called www.NoCarbonTaxes.com, with a petition to oppose any form of carbon tax. The CTF petition pointed out that carbon taxes make everyone poorer and do not help household budgets or boost the economy.

Canadian voters didn't like Dion's carbon tax, either. In the 2008 election, the Conservatives formed government, while the Liberals lost eighteen seats and received twenty-six percent of the popular vote—the worst popular vote result for the Liberal Party in more than a century. The NDP, with Layton hammering away at the carbon tax, gained seven seats.

Columnist Andrew Coyne highlighted how pushing a carbon tax hurt the Liberals.

"The Liberals have been steadily losing altitude throughout the campaign ... the Green Shift/carbon tax has by all accounts been a major contributor," Coyne wrote in October 2008.

Polling close to election day showed that a majority of Canadians opposed Dion's plan. The NDP held strong on opposing Dion's carbon tax during the election despite pressure from within their own tent. "To oppose [the carbon tax plan], it's just nonsense," said David Suzuki. "It's certainly the way we got to go."

"The Liberals ran third, behind the NDP, in every Western province," said MacKinnon. "From Manitoba to British Columbia, Western Canadian voters rejected the Liberal Party in record numbers. As the Liberals assess the last election and look to the future, they should understand that the Green Shift plan reinforced many Westerners' view that the Liberal Party is an eastern based entity that does not understand Western Canada and cannot be trusted to protect its interests ... An important first step for the Liberals will be to clarify their position on a federal carbon tax."

The next Liberal leader got the message.

"You can't win elections if you're adding to the input costs of a farmer putting diesel into his tractor, or you're adding to the input cost of a fisherman putting diesel into his fishing boat, or a trucker transporting goods," said Michael Ignatieff. "You've got to work with the grain of Canadians and not against them. I think we learned a lesson in the last election."

But the lesson didn't stick. In 2011, the Liberal platform vaguely suggested it would implement a cap-and-trade system and engage the American government on key environmental issues, including "carbon pricing."

Unlike the Liberals, Layton's NDP maintained its opposition to a consumer carbon tax and it paid off. In 2011, the NDP captured 103 seats during Layton's Orange Wave, forming official opposition for the first time in party history.

"The well-documented reality is, the NDP has opposed a carbon tax in the past and continues to do so now," the NDP said in 2012.

Following the 2011 election, the Conservatives stopped flirting with carbon taxes altogether, at least until Erin O'Toole became party leader roughly a decade later. The 2015 Conservative election platform was explicitly against implementing any form of carbon tax, stating the "solution to the climate change challenge must come from innovation ... not by closing down our vital natural resources industries or imposing job-killing carbon taxes."

But in 2015, a federal carbon tax once again reared its ugly head, with Trudeau's vague notion of implementing a "price on pollution." And in 2019, that vague notion became reality, when he imposed his national carbon tax on Canadians.

CHAPTER 3

CARBON TAX CONSENSUS

To understand how far the carbon tax fell, look back at the peak of its popularity, when it climbed to the highest of political heights: consensus.

"We recognize that the most efficient way to reduce our emissions is to use pricing mechanisms," said Conservative leader Erin O'Toole on April 15, 2021. "However, having a market-based approach means that we cannot ignore the fact that our largest and most integrated trading partner—the United States—does not yet have a national carbon pricing system."

Then O'Toole added: "Not a cent goes to Ottawa."

With that announcement, O'Toole made it unanimous: every major political party would support a carbon tax in the 2021 federal election.

Under O'Toole's plan, Canadians would be hit with the carbon tax when they bought gas or paid for their home heating, just like they were under Liberal Prime Minister Justin Trudeau's carbon tax. But instead of rebate cheques, O'Toole's carbon tax would see the money put into bank accounts Canadians could use for government-approved, environmentally friendly purchases. He also added a proposed tariff on imports from "bad actors" like China.

But even at that peak of popularity in 2021, the problems with carbon taxes, no matter which party imposes them, were already apparent.

"Pricing mechanisms" is just another euphemism for taxes. And Canadians are well aware that taxes make life more expensive.

Most of the world, including the United States, does not have a national carbon tax. That makes punishing Canadians with one environmentally symbolic, but practically ineffective.

And O'Toole's insistence that Ottawa wouldn't keep a penny of the revenue money generated from his carbon tax highlighted an important reality: the entire carbon tax scheme depended on convincing Canadians the government could take their money and run it through the bureaucratic machinery, only to turn around and give it all (and more) back, thereby making them better off financially.

This central problem would loom large in the coming years, but in the meantime, O'Toole also had a procedural problem to deal with. He didn't build support for his position. Instead, he lied about it and then scrambled to sell his surprise carbon tax to Canadians.

O'Toole wasn't the first politician to use this tactic, but he took it to the extreme. Over and over, politicians would impose carbon taxes on Canadians without campaigning on them. As a result, these politicians, from all political parties, pretended to manufacture consent for carbon taxes without actually getting Canadians on board.

To make matters worse for O'Toole, his betrayal was in black and white.

During the Conservative leadership race, O'Toole joined all other candidates in signing a Canadian Taxpayers Federation (CTF) pledge.

"I, Erin O'Toole, promise that, if elected prime minister of Canada, I will: Immediately repeal the Trudeau carbon tax; and reject any future national carbon tax or cap-and-trade scheme."

But a few months after winning the Conservative Party leadership race in 2020, the CTF got a tip that O'Toole would endorse a form of carbon tax embedded in fuel regulations.

The CTF readied news releases and newspaper columns blasting O'Toole for his potential betrayal. But O'Toole's office swore the rumours were false.

The CTF kept checking.

CTF president and CEO Scott Hennig called member of Parliament and Conservative caucus chair Tom Kmiec to ask about the rumours. Kmiec was stunned by the rumour, but he double checked with O'Toole personally, as well as with his chief of staff. He called Hennig back with a reaffirmation: Both O'Toole and his top advisor confirmed the rumour was false.

The CTF attacks went unsent.

O'Toole was absolutely against the carbon tax. That is, until he was for it on April 15, 2021.

Pro-carbon tax advocates celebrated.

"The battles have been won, and the old political axioms are dead—Carbon taxes are here to stay," declared an opinion piece in the *Globe and Mail*.

The headline lacked situational awareness. And as a former member of the Royal Canadian Air Force, O'Toole should have been watching his six.

"Erin O'Toole faces backlash from 'surprised and frustrated' Conservatives over carbon pricing plan," proclaimed a *National Post* headline days after O'Toole's heel turn.

"I think a lot of people are surprised and frustrated," an anonymous MP told the *Post*.

MPs were more explicit in conversations with the CTF.

Kmiec called Hennig. He was livid about the flip-flop. O'Toole hadn't even mentioned his carbon tax during the caucus meeting the day before the announcement. Conservative MPs were outraged. Kmiec was personally angry that his leader had used him to falsely deny the earlier rumours.

Another Conservative MP, John Williamson, reached out to Hennig to assure him nobody in caucus saw the betrayal coming and would have mobilized to stop it if they had.

An analysis in the *National Post* studied the social media posts of Conservative MPs. Usually, politicians repost whatever their leaders say. A few Conservatives applauded O'Toole's carbon tax, but most were silent.

And the *Post* highlighted one presciently quiet reaction to O'Toole's carbon tax reversal: "Pierre Poilievre, with 134,500 followers, made no mention of it."

What political elites like O'Toole missed was bubbling right below the surface.

Canadians were frustrated with high levels of taxation, even before inflation rose to a forty-year high in 2022. This sentiment was exacerbated by a pandemic that rocked the private sector while those sheltered behind the golden gates of government continued to enjoy job security and annual bonuses and pay raises.

Political talking points couldn't convince Canadians to ignore what they saw on their gas-station receipts. They didn't buy the idea that a carbon tax could make their life more affordable. A tax on gasoline and diesel, on natural gas and propane, on truckers and farmers, is a tax on the necessities of life in Canada. And as inflation continued to rise, so, too, did resentment toward the carbon tax.

Far too many Canadians saw the carbon tax for what it was: an attempt by the elites in the government, the universities, the media and the activist class to punish lifestyles they abhorred and dictate to Canadians how they should live.

In other words: do what we say—or pay.

O'Toole's carbon tax gambit failed. He lost the 2021 federal election and soon enough his own caucus repaid his carbon tax betrayal with knives in his back.

In the subsequent Conservative Party leadership race, every candidate opposed the carbon tax. Poilievre won in a runaway.

Soon, opposition to the carbon tax was everywhere. Conservative crowds chanted "axe the tax" at rallies. Conservative, Liberal and NDP premiers came out in opposition too. Even Liberal stalwarts began to admit the carbon tax had become deeply unpopular and might have to be repealed.

Here's the reality: the carbon tax was deeply flawed from the beginning.

And the burgeoning political consensus pushing back against the tax was a harbinger of the revolt to come.

CHAPTER 4

CONSERVATIVE PARTY REVOLT

Erin O'Toole's flip-flop seemed like a death blow to carbon tax opponents. It seemed like there was "no longer a viable political coalition to be built around fighting carbon taxes," according to authors in the *Globe and Mail*. It seemed like "carbon taxes [were] here to stay."

But things aren't always what they seem.

The revolt began with grassroots opposition to O'Toole's flip-flop, which eventually led to Conservative politicians organizing against their party leader.

Minutes after O'Toole broke his promise to fight the carbon tax, the Canadian Taxpayers Federation (CTF) launched an aggressive campaign to hold him accountable.

The same morning, the CTF put out a media statement slamming "CPC Leader Erin O'Toole for announcing his own carbon tax." The press release quoted the pledge O'Toole signed less than a year earlier to fight carbon taxes.

The pledge O'Toole made to taxpayers was unequivocal: "I, Erin O'Toole, promise that, if elected prime minister of Canada, I will: Immediately repeal the Trudeau carbon tax; and reject any future national carbon tax or cap-and-trade scheme."

The CTF did dozens of interviews on TV and radio reminding Canadians that O'Toole had sold them out on the carbon tax.

The CTF even appeared on CTV's *Question Period* right after O'Toole's environment critic. O'Toole had sent him on air to mop up his mess and defend the Conservative's carbon tax disaster. The CTF reminded all those watching of O'Toole's broken promise, and made clear that carbon taxes—regardless of which party imposes them—make the necessities of life more expensive.

One month after O'Toole's flip-flop, the CTF released its annual Gas Tax Honesty report that highlights all the taxes Canadians pay on fuel. But that year's Gas Tax Honesty press conference was a little different.

The CTF had a special guest: Fibber, the long-nosed honesty in politics mascot, who bears a striking resemblance to Pinocchio. Fibber was there to help us spotlight broken carbon tax promises from Prime Minister Justin Trudeau and O'Toole.

Fibber held up a big sign with a quote from Trudeau's first environment minister, Catherine McKenna: "The commitment was to go up to 2022. There was no intention to go up beyond that, there's no secret agenda."

Of course, following McKenna's statement, Trudeau announced he would keep hiking his carbon tax every year until 2030—and perhaps beyond that.

Alongside Fibber, I held up a big sign with O'Toole's broken promise to "repeal the Trudeau carbon tax" and "reject any future national carbon tax or cap-and-trade scheme."

Thousands of CTF supporters emailed Conservative MPs to blast them over O'Toole's support for carbon taxes. Those very same supporters continued to give MPs an earful as they knocked on doors during the 2021 election.

Videos and graphics flashed all over social media railing against O'Toole's carbon tax flip-flop. One memorable social media graphic was a dollar with O'Toole's face on it, mocking his carbon tax rewards program. If implemented, O'Toole's carbon tax would raise the price of gas, diesel and home heating fuels, then return money to Canadians through a Low Carbon Savings Account. The money could then be spent on government-approved green goodies.

The Trudeau carbon tax is a thinly veiled redistribution scheme, but at least some Canadians got actual cash back in their bank accounts. Under O'Toole's plan, Canadians would get money in their Low Carbon Savings Account, which they could spent on so-called green products on a government-approved list. This created a question for Canadians: cash back under the Trudeau model, or O'Toole bucks toward buying a solar-powered e-bike?

Perhaps the most memorable part of this campaign were the billboards. They showed a big picture of O'Toole smiling beside the anti-carbon tax pledge he signed and later broke. On the ad was a simple message to O'Toole and his Conservative MPs: "Keep your promise. Scrap the carbon tax!" (In fact, if you go to Google Maps, type in the address for the prime minister's office, and zoom in on the bus shelter across the street, you can still see the CTF ad with Parliament Hill in the background.)

Conservative politicians got the message. Without a doubt, they certainly got the message from voters in the 2021 election.

O'Toole and his campaign strategists thought that by giving in on the carbon tax, they would increase the Conservative's chances to form government. But the election results proved O'Toole's carbon tax flip-flop was a political loser.

In 2019, when the Conservatives ran against the carbon tax, they received 34.3 percent of the popular vote. In 2021, under O'Toole, the Conservatives received 33.7 percent of the popular vote. In 2019, 121 Conservative MPs were elected. In 2021, the Conservatives lost two seats, with 119 MPs elected. Under O'Toole and his carbon tax, the Conservatives lost seats in the House of Commons and lowered its share of the popular vote.

While elections depend on many factors, one reason the Conservatives did so poorly is they weren't able to wedge the Trudeau

Liberals on affordability. The election spanned August and September 2021. The spring of 2021 marked the beginning of Canada's inflation crisis.

In typical times, inflation sits somewhere around two percent. In August 2021, inflation was double that, at 4.1 percent. A large contributor to the higher cost of living was the carbon tax pushing up gasoline prices.

"Year over year, gasoline prices rose 32.5 percent in August [2021]," according to Statistics Canada. In September 2021, inflation reached 4.4 percent, "with transportation prices (+9.1 percent) contributing the most to the all-items increase."

Prices had been rising above the typical two percent in the lead up to Trudeau announcing the election in August 2021. Inflation was 3.7 percent in July, 3.1 percent in June, 3.6 percent in May and 3.4 percent in April, according to Statistics Canada.

An Abacus Data poll released one day before the final votes were cast in the 2021 election showed the top issue among voters, by a wide margin, was the cost of living.

"When we ask respondents to pick the top two issues that will determine their vote, the most frequently chosen is reducing the cost of living (thirty-six percent)," according to Abacus Data. The second most pressing issue was health care, eleven percentage points below the cost of living.

In fact, among voters who supported each of the three top national parties—the Conservatives, the Liberals and the New Democrats—the number one issue was the cost of living. This makes clear that fighting the carbon tax as inflation was spiralling out of control would have been a political winner for the Conservatives had O'Toole not flip-flopped.

Ultimately, it was Canadians who held O'Toole accountable. And after being berated by Canadians at the door steps over the Conservative Party's carbon tax flip-flop, so did Conservative MPs.

In January 2022, while O'Toole was still leader and the carbon tax was still official policy, the CTF obtained pamphlets Conservative MPs were handing out to their constituents. And they didn't mince words about the carbon tax.

"We will not support Justin Trudeau's inflation agenda," reads a letter Conservative MP Arnold Viersen sent to his constituents in northern Alberta. "My Conservative colleagues and I are calling on the Liberals to immediately repeal the carbon tax."

MP Viersen told the online news site *True North* he opposed carbon taxes, even if imposed by his own party.

"I've been clear from the onset of the imposition of the carbon tax that it only makes life more expensive for people in northern Alberta and doesn't meet any of its goals," Viersen said. "I've been having a lot of folks send me copies of their natural gas bill. December, for most people, has been the highest natural gas bills they've ever had in their entire lives. I will be opposed to any tax that makes life more expensive, put forward by any leader in the federal Parliament."

That was a clear shot at both Trudeau and O'Toole. And soon enough, the shots from Conservatives toward their own leader became louder and more direct.

Conservative MP Bob Benzen cited the carbon tax in his letter calling for a leadership review for O'Toole's on January 31, 2022.

"The adoption of a de-facto carbon tax policy in April 2021 despite clear direction from our members who are opposed to a carbon tax and despite his campaigning against such a tax during the leadership contest" topped Benzen's list of grievances against O'Toole.

The Battlefords-Lloydminster Conservative Electoral District Association launched a website advocating to "end the carbon tax."

What had happened was clear. O'Toole sold Canadians out on the carbon tax. Canadians put pressure on O'Toole's MPs. And that pressure forced those MPs to oust O'Toole.

An internal Conservative caucus revolt removed O'Toole as leader on February 2, 2022. Just one week later, Interim Conservative Party Leader Candice Bergen announced the party had dropped its support for carbon taxes.

"We believe that there should be no federally imposed carbon taxes or cap-and-trade systems on either the provinces or on the citizens of Canada," the Conservative policy book stated. "The provinces and territories should be free to develop their own climate change policies, without federal interference or federal penalties or incentives."

And just like that, the carbon tax was front and centre in the political debate again.

MP Pierre Poilievre announced he was running to be the new leader of the Conservative Party on February 5, 2022. He quickly emerged as a front-runner and made axing the carbon tax the central plank of his campaign.

The CTF interviewed all candidates on its podcast. Here's what we asked Poilievre about the carbon tax: "When are you going to scrap the carbon tax, before or after lunch on your first day if your elected prime minister?"

Poilievre responded: "It will be in my very first budget, that's for sure. Normally budgets are introduced after lunch, so it will be a nice dessert."

Given O'Toole's flip-flop, we had to ask Poilievre a follow-up question: "Will you replace the Trudeau carbon tax with any form of carbon tax of your own and what are you going to do with that second carbon tax the government is bringing in through fuel regulations?"

Poilievre's answer: "I will not bring in any other carbon tax and I'm against the second carbon tax as well."

In fact, every leadership candidate confirmed they would scrap Trudeau's consumer carbon tax that drives up prices at the gas pumps and on home heating bills.

Poilievre won the Conservative Party leadership contest with more than seventy percent of the vote. Ever since his victory, Poilievre made Axe the Tax his signature promise to voters. And you'd be hard pressed to find a day in the House of Commons where Poilievre hasn't pushed for a "carbon tax election." He has continued to beat the anti-carbon tax drum at every opportunity and helped keep the issue front and centre in Canadian politics.

Poilievre deserves credit for fighting the carbon tax harder than any federal politician in Canadian history. Previous federal carbon tax critics flirted with different forms of carbon taxes or were often muted in their criticism. Poilievre was unequivocal and on the offensive: axe the tax. And Canadians appreciated his hard stance.

"Axing the tax—it means that I have a chance, that there's a chance that my family and I are going to survive," Sarah Morin, a forty-one-year-old stay-at-home mother of two, told the *Canadian Press*.

Poilievre disproved the central thesis of the *Globe and Mail* article that claimed there was "no longer a viable political coalition to be built around fighting carbon taxes." In fact, he brought the Conservative Party to its highest polling numbers in decades with his Axe the Tax campaign.

It turns out the carbon tax wasn't here to stay.

CHAPTER 5

PROVINCIAL CARBON TAXES IN CANADA

British Columbia was the first province to go all in on the carbon tax experiment in 2008 and the promised pay-offs were huge: big tax cuts that would make the carbon tax revenue neutral and big drops in emissions.

The *New York Times* said it "set British Columbia apart as a leader on the cutting edge."

The Times went head over heels for the policy, publishing several opinion pieces praising BC's carbon tax. In a 2016 editorial, the paper noted its "real value may lie in providing a template for the rest of the world." Various writers would go on to use the BC carbon tax as a model for similar policies they wanted to see adopted in the US.

"We can move past the partisan fireworks over global warming by turning British Columbia's carbon tax into a made-in-America solution," wrote Yoram Bauman, an environmental economist, and Shi-Ling Hsu, a law professor at Florida State University.

Referring to BC's carbon tax, *The Economist* said, "We have a winner."

Thomas Pedersen, former director of the Pacific Institute for Climate Solutions and author of *The Carbon Tax Question: Clarifying Canada's Most Consequential Policy Debate*, called BC's carbon tax a "template for the world." Both the United Nations and the World Bank declared BC's carbon tax the model to follow. And the OECD called it a "textbook" example of how to implement climate policy.

While the government of British Columbia had buy-in from domestic and international elites, alike, it never had buy-in from the one group that truly matters: the people.

The BC carbon tax was imposed by the Liberal government of Premier Gordon Campbell. It's worth noting the Campbell Liberal's 2005 election platform made no mention of a carbon tax. In fact, the platform didn't even include the word "carbon." And the only mention of an environmental tax referenced plans to *cut* taxes: "Continue new tax reductions for alternative fuel vehicles."

When the Campbell government brought in its carbon tax, it made two fundamental promises to British Columbians: to dramatically reduce emissions and return all carbon tax revenue to taxpayers through offsetting tax cuts. But those promises soon became casualties of reality.

"The carbon tax is one of several key building blocks to help government reduce BC's greenhouse gas emissions by thirty-three percent below 2007 levels by 2020," the government said in its 2008 budget.

The promise couldn't have been clearer: by 2020, the carbon tax would result in BC's emissions declining by thirty-three percent below 2007 levels.

The result? By 2020, right before the onset of the COVID-19 pandemic, BC's emissions had *increased* by one percent over 2007 levels, according to the government's emissions data.

The BC government also promised its carbon tax would be "revenue neutral." That meant "all revenue generated by the carbon tax" would be "returned to individuals and businesses through reductions to other taxes." The government said it would "require a plan to be tabled in the legislature each year, showing how the revenue raised will be returned to taxpayers," and that "none of the carbon tax revenue will be used for expenditure programs."

The provincial government's 2008 budget attempted to offset the cost of its carbon tax by cutting the bottom two personal income tax rates and reducing the corporate income tax rate and the small business tax rate by one percentage point. The government also introduced a low-income climate action refundable tax credit.

The *New York Times* praised the move, noting a "big appeal of [BC's] system is that it is essentially revenue-neutral." Another *New York Times* column proclaimed that carbon tax hikes were "good news not only for the environment but for nearly everyone who pays taxes in British Columbia, because the carbon tax is used to reduce taxes for individuals and businesses."

If the government was going to impose a carbon tax that made the necessities of life more expensive, at least it was cutting other taxes. The only problem? This "revenue neutral" tax swap was a wolf in sheep's clothing. Shortly after the government imposed the carbon tax, it quickly proved not to be revenue neutral at all.

By 2013, the government was no longer solely relying on new tax measures to offset carbon tax revenue. Instead, it used pre-existing tax reductions in its revenue-neutral calculations. In other words, the government was relying on previously introduced tax cuts to claim its carbon tax was returning as much money back to taxpayers as it took. In fact, some of the tax credits the government counted as offsets were introduced in the 1990s.

"If the pre-existing tax measures are properly removed from the government's revenue neutral calculation, then BC's carbon tax ceases to be revenue neutral as of 2013–14, with a net tax increase of $226 million that year," according to the Fraser Institute, a non-partisan, public policy think tank. "From 2013–14 to 2018–19, the carbon tax is projected to result in a cumulative $865 million net tax increase for British Columbians. If we were to distribute this tax increase equally among the province's populace, each British Columbian would pay $182 more per person, or $728 for a family of four."

Far from being revenue neutral, the BC carbon tax started to cost provincial taxpayers hundreds of millions of dollars annually just five short years after it was implemented. And while the government initially cut taxes broadly to offset the cost of the carbon tax, it soon turned to boutique tax credits.

"Just five years later, as the carbon tax revenue increased, the government no longer provided new tax cuts that sufficiently offset the carbon tax's revenue," wrote the Fraser Institute. "Despite what the government claims, BC's carbon tax is not actually revenue neutral."

BC's experiment with a carbon tax was a failure on all fronts.

The government did not run on a carbon tax before it imposed one, meaning taxpayers weren't given a voice on this consequential issue. It promised emissions would decline by 2020, but instead they increased. It also promised all the money the carbon tax took from taxpayers would be offset through tax cuts, but five years later it broke that promise.

As the Fraser Institute rightly concluded, "BC's carbon tax is not the 'gold standard' it's often made out to be."

* * *

When Prime Minister Justin Trudeau announced his intention to impose a country-wide carbon tax on Canadians in October 2016, things looked bleak for taxpayers.

BC's carbon tax had been in place for nearly a decade. Quebec had already imposed its cap-and-trade carbon tax. The governments of Ontario and Alberta were about to impose carbon taxes of their own. And the political leadership in the rest of the Canadian provinces were either silent or supportive of carbon taxes.

At that time, there were only two forces in Canada publicly opposing the carbon tax. One was the Canadian Taxpayers Federation (CTF). The other was former Saskatchewan premier Brad Wall and his government.

"The national focus on carbon pricing holds the lowest potential for reducing emissions, while potentially doing the greatest harm to the Canadian economy," Wall said in 2016. "We produce less than two percent of global [greenhouse gas] emissions. Whatever impact the federal carbon tax will have on Canada's emissions, global GHG emissions will continue to rise because of the developing world's reliance on coal-fired electricity."

Wall promised to "investigate all options to mitigate the impact of one of the largest national tax increases in Canadian history." It was clear Wall knew fighting the carbon tax would be a tough, lonely battle. "I think we'll be on our own right now," Wall told the CBC.

But Wall's willingness to be the lone government in the wilderness speaking out for taxpayers would soon ignite opposition against the carbon tax from other provinces.

* * *

While Alberta is now considered the province leading the crusade against carbon taxes, that wasn't always the case.

Premier Rachel Notley's NDP government had a surprise for Wild Rose country in 2017: a carbon tax.

The Alberta NDP ran on a twenty-five-page platform in 2015. There was no mention of a carbon tax in the platform and nothing that even insinuated one was coming. The province's politicians never bothered asking voters whether they supported a carbon tax. Instead, they implemented their new revenue tool after being elected and completely circumvented the people.

It's easy to understand why the NDP chose not to run on a carbon tax. Polls showed Albertans did not support one. About two months before Alberta's carbon tax came into effect, the CBC reported on polling that showed sixty-seven percent of Albertans opposed the policy. Two months after the carbon tax was implemented, Postmedia released polling by Mainstreet Research showing nearly two-thirds of Albertans remained opposed. In 2018, Mainstreet released another poll showing sixty-five percent of Albertans were against the carbon tax.

When Albertans finally got a chance to vote on the carbon tax during the 2019 provincial election, they sent the government who imposed it packing.

Alberta's carbon tax was never voted on, never received public buy-in, and hammered families and businesses during tough economic times. And just like in BC, Alberta's politicians falsely promised the policy would be revenue neutral.

The people were relegated to the background, as big business and big government congratulated each other over the carbon tax. Even big oil and gas companies like Suncor, Shell, Cenovus and Canadian Natural Resources came out to support the carbon tax at Notley's press conference introducing it.

"We think it is literally a game changer," said Suncor president and CEO Steve Williams, adding it was a "historic day" for Alberta's oil sands.

"I would say the most important part of the announcement … was probably who was on stage with the premier," said University of Alberta economist Andrew Leach, who headed up Notley's task force on climate change.

The CTF countered the apparent consensus with careful analysis of the Alberta carbon tax.

The first step in the fight was showing Albertans how much the carbon tax would cost them.

"Alberta's new broad-based carbon tax is a tax on everything that moves," wrote Paige MacPherson, then the CTF's Alberta director, in the *Financial Post*. "It will raise the price of gas and home heating, costing Alberta families an estimated $300 to $600 per year. That will rise to over $900 per year for the average family by 2030. The price of clothing, food and everything transported will also increase … The province will lose investment and rural communities will be hit hard."

The CTF launched a Scrap the Tax campaign, which included billboards across the province.

"By 2018, the carbon tax will cost your family over $600/per year," the billboards read.

The pressure campaign included Scrap the Carbon Tax bumper stickers, tens of thousands of petition signatures, hundreds of media interviews and viral social media videos.

The CTF even flew in an Australian economist to detail his country's successful fight to abolish its carbon tax in 2014. Chris Berg, a senior fellow with Australia's Institute of Public Affairs, met with media and businesses, as well as opposition politicians in Alberta, to talk about why his country repealed its carbon tax within a few short years of implementing it.

"It was a Down Under blunder that should serve as a lesson to Alberta as it embarks on its own controversial carbon tax," wrote the *Calgary Herald*, after meeting with Berg and the CTF.

"We were told we would lead the world, but it didn't look like the world was interested in carbon taxes," Berg said. "At least four party political leaders lost their job over the issue. It cost the Australian economy $8 billion a year for two years, it raised electricity prices by twenty-five percent … and contributed to higher prices at the supermarket. It's easy to say we're making big polluters pay, but fundamentally these taxes have an effect on individual people—people have to pay these taxes."

In addition to using traditional and social media, along with everyday Albertans, to push the NDP government to scrap the carbon tax, the CTF also had to put pressure on the province's most conservative political party, the Wildrose. In the early days, the Wildrose's position on carbon taxes could be best described as murky.

When the carbon tax was first announced in November 2015, official opposition leader Brian Jean came out swinging against the tax.

"Why is the premier more determined to create another new tax on Albertans rather than working to protect Albertans' jobs?" Jean said.

Jean dubbed the carbon tax a "tax on and everything" and noted it was "the latest blow to Albertans who are already losing their jobs or seeing their take-home pay cut." He also acknowledged the tax grab was "not going to reduce emissions."

But soon the Wildrose's tune changed.

Jean still vowed to kill Notley's carbon tax, but his stance had softened by early 2016, saying, "I can't tell you that the Wildrose wouldn't bring in a carbon tax in the future." In fact, the *Calgary Herald* reported the Wildrose "would consider replacing [the NDP's carbon tax] with a revenue-neutral levy" if it formed government.

In March 2016, the Wildrose released its Jobs Action Plan, which included a promise to "cancel the NDP's $3 billion carbon tax until a full economic impact analysis is conducted." The Wildrose was not signalling its complete opposition to carbon taxes, only its opposition to the NDP's carbon tax, as well as a promise for further study.

Albertans knew where the NDP stood: the New Democrats loved carbon taxes. The real question was: Did the Wildrose want to kill the carbon tax, did it want to review it, or did it just want to replace it with another carbon tax of its own? Even Notley was skeptical Jean would scrap her carbon tax.

"I actually don't even agree that he would do it," Notley told the *Calgary Herald*.

The CTF turned up the pressure on Alberta's Official Opposition with media releases leveraging grassroots opposition to carbon taxes, pressuring Jean to take a clear stance.

"Albertans and rural residents need answers from Wildrose. Carbon tax: yes or no?" said MacPherson in March 2016. "If elected government, the Wildrose Party should do the right thing for taxpayers, be clear and re-commit to scrapping any carbon tax in Alberta, full stop."

Months of pressure from the CTF worked. In October 2016, Wildrose members overwhelmingly voted at the party's annual general meeting to oppose carbon taxes and repeal the provincial carbon tax if the party formed government.

Suddenly, Jean rediscovered his opposition to the carbon tax.

"The members have ratified exactly our position on a very bad and regressive tax," Jean said, while also calling on the government to hold a referendum on the carbon tax.

The next turning point in Albertans' fight against the carbon tax was when former MP (and former head of the CTF) Jason Kenney announced his plan to "unite the right" in the province and throw his hat in the ring to be the next premier.

In 2017, the Progressive Conservatives and the Wildrose agreed to merge and form the United Conservative Party (UCP).

The CTF had the three leading UCP leadership candidates—Kenney, Jean, and Doug Schweitzer—sign a pledge to scrap Alberta's carbon tax "within 100 days of taking office" if they became premier.

Kenney won the UCP leadership on the first ballot with sixty-one percent of the vote. His promise to Albertans was unequivocal.

"An Alberta United Conservative government will immediately convene the provincial legislature to pass bill No. 1, the carbon tax repeal act," Kenney said. "If they seek to impose an Ottawa tax on us, we will sue Justin Trudeau."

And Kenney did just that.

The United Conservative Party swept the NDP from power in the 2019 election and the carbon tax was a key issue of the campaign.

"The NDP introduced the largest tax increase in Alberta history without campaigning on it," the UCP platform read.

After forming government, Kenney scrapped the Alberta carbon tax with his very first bill. And less than a month later, Kenney launched a constitutional court challenge against the Trudeau government's carbon tax, which would be imposed on Albertans the following year.

* * *

Liberal premier Kathleen Wynne imposed a cap-and-trade scheme in Ontario, continuing the pattern of provinces implementing carbon taxes as a post-election surprise.

Wynne had not campaigned on a carbon tax during the 2014 Ontario election. The tax was announced in 2015 and came into effect in 2017, initially increasing the price of gas by 4.3 cents a litre and the price of diesel by about five cents a litre. It also increased natural gas bills by $60 per year.

"Call it carbon pricing, cap and trade, a market mechanism or—I believe it's misleading—if you must, go ahead and call it a tax," Wynne said.

It looked like Ontario would be saddled with a carbon tax for the foreseeable future. At that time, Ontario's Progressive Conservative Party, the official opposition, didn't fight carbon taxes like it does today. In a speech to the PC convention in 2016, party leader Patrick Brown said he was in favour of carbon taxes.

Talking about climate change, Brown said, "We have to do something about it and that something includes putting a price on carbon."

Brown resigned as party leader in 2018. Ontario was poised to have a provincial election in four months' time. The Ontario PCs needed a new leader. Four candidates ran for the top job: Doug Ford, Christine Elliot, Caroline Mulroney, and Tanya Granic-Allen.

It was unclear if the new party leader would follow in Brown's footsteps and support a carbon tax in Ontario. The CTF's Ontario director at the time, Christine Van Geyn, quickly cleared up that question by getting all four leadership candidates, including eventual winner Doug Ford, to sign a "no carbon tax" pledge.

After winning his leadership bid, Ford handily won the provincial election and became premier of Ontario in the summer of 2018. The formerly governing Liberals faced their worst defeat in party history and the carbon tax was a key election issue.

Ford made good on his pledge and passed The Cap and Trade Cancellation Act, scrapping Ontario's provincial carbon tax in 2018, saving taxpayers $7.2 billion. Ford also announced he would be taking the federal government to court over its plan to impose a carbon tax across the country. He even went so far as to call Trudeau's carbon tax "the worst tax ever."

"The federal government's carbon tax is forcing Ontarians to pay more to heat their homes, drive to work and buy groceries," said former attorney general Caroline Mulroney, when announcing Ontario's constitutional challenge. "It's simply not fair to hardworking individuals, families and small businesses."

The Ontario government argued, "the provinces, not the federal government, have the primary responsibility to regulate greenhouse gas emissions and that the charges the act imposes are in fact unconstitutional disguised taxation."

Back then, some of the judges and those in the federal government were still trying to paint the carbon tax as a "levy," which according to them was something altogether different than a tax, despite costing just as much.

Ontario lost the initial case with Saskatchewan and New Brunswick. The case was then appealed to the Supreme Court of Canada, and the provinces of Manitoba, Alberta and Quebec joined the fight.

Opposition to the carbon tax was growing.

* * *

Nobody came out of the carbon tax fight looking sillier than former Manitoba premier Brian Pallister.

"Mr. Pallister has flip-flopped on this issue more times than a pickerel on a dock," said NDP Leader Wab Kinew in 2020, before he was elected Manitoba premier.

Pallister started off by making the typical mistake: he surprised taxpayers with a carbon tax that he hadn't campaigned on.

Manitoba's Progressive Conservatives won the 2016 election in large part because the incumbent NDP government had promised not to raise the provincial sales tax but broke that promise and raised it anyway. Taxpayers were unimpressed and Pallister capitalized on that anger to win the election.

He didn't campaign on a carbon tax. In fact, the forty-four-page platform said only that his government would develop a policy for "carbon pricing that fosters emissions reduction, keeps investment capital here and stimulates new innovation in clean energy, businesses and jobs."

That's it. No mention of the billions of dollars it would cost Manitobans, just a vague policy note buried in a platform.

Pallister's carbon tax rollout went from typical to ridiculous with his government's 2018 budget.

"It's the largest tax cut in Manitoba history," said Finance Minister Cameron Friesen in his budget speech.

The press release highlighted an income tax cut that would save Manitobans about $156 million that year.

But the actual budget documents told a different story: Pallister was implementing a carbon tax that would *cost* taxpayers about $248 million. In other words, Pallister's "historic" tax cut was actually a net tax hike after accounting for the carbon tax he hadn't campaigned on.

Taxpayers were furious, but they weren't the only ones panning Pallister's carbon tax policy.

Prime Minister Justin Trudeau told CBC he was happy to see a "leader, indeed a conservative leader, who understands the need to have a concrete plan."

But the prime minister went on to say the conversation would have to continue to make sure "everyone's doing their part."

Pallister saw the threat behind the smiling veil.

"I want respect from the federal government in respect of our plan," said Pallister. "To have the federal government threaten to … double up on our carbon tax is not helpful."

Pallister was proposing to impose a carbon tax of 5.3 cents per litre of gasoline ($25 per tonne of emissions). But he promised to keep his carbon tax at that level rather than doubling it as required by the federal backstop at the time.

Predictably, the Trudeau government rejected Pallister's flat carbon tax proposal and imposed the federal carbon tax.

The Manitoba Progressive Conservatives explicitly opposed the carbon tax during the 2019 election.

"PC government will move forward with its Made-in-Manitoba Climate and Green Plan, without a carbon tax," stated a Progressive Conservative petition.

But there was still more flopping left in that pickerel on the dock.

Pallister re-introduced his flat carbon tax in the 2020 budget after an Alberta court ruled against the federal carbon tax backstop. This time he tried to sweeten the betrayal with a sales tax cut.

But then the pandemic hit before Manitoba could implement the provincial carbon tax and Pallister hit pause on its re-implementation.

And that's when the flopping stopped, and Manitoba cut the line on a provincial carbon tax. Heather Stefanson won the Conservative leadership race to replace Pallister and left the carbon tax untouched during her brief time as premier.

Wab Kinew's NDP promised to suspend the provincial gas tax entirely and won the 2023 Manitoba election.

The prime minister went from having a Conservative premier proposing his own carbon tax to a NDP premier actively opposing the scheme.

Kinew told *CTV* that Manitoba doesn't "need that carbon tax here" at the end of 2024.

"Give Manitobans a break," Kinew said. "We've seen what a difference it's made for your budget on the provincial level to get rid of the fuel tax. Imagine what would happen if that carbon tax got taken off to help you and your family."

* * *

Saskatchewan consistently opposed the carbon tax even when Premier Scott Moe took over from Brad Wall.

In October 2016, Moe, as environment minister, had attended a meeting in Montreal with his federal counterpart, Catherine McKenna. The meeting had been called so McKenna could hash out an agreement on carbon taxes with the provinces.

While the meeting was underway, Trudeau rose in the House of Commons and announced his government would unilaterally impose a federal carbon tax on any province that refused to implement a carbon tax of its own.

As a result of the announcement, Moe walked out of the meeting.

"The level of disrespect shown by the prime minister and his government today is stunning," Wall said. "This is a betrayal of the statements made by the prime minister in Vancouver this March. And this new tax will damage our economy."

"Many Westerners will see this as National Energy Program 2.0," Moe said, referencing the divisive, anti-West policy implemented by Trudeau's father. "It's not a good day for federal-provincial relations."

Maclean's magazine ran a 2018 cover story entitled "A carbon tax? Just try them."

The picture included federal Conservative Party leader Andrew Scheer, surrounded by Premiers Wall, Ford and Pallister, as well as Alberta opposition leader Kenney. On the cover in bold letters were the words: "The resistance."

Columnist Paul Wells concluded the piece by noting Trudeau "will be tested" on the carbon tax "not only by Scheer, but by all the Conservative leader's provincial allies."

Wells was spot on.

Initial political resistance to Trudeau's carbon tax started in Saskatchewan with Wall and then Moe. The resistance grew and spread to Ontario and Ford, followed by Alberta and Kenney. The resistance grew so large that it eventually erupted into court challenges, with the provinces taking the federal government to the Supreme Court over Trudeau's carbon tax. Without Wall holding the line in the early days of the Trudeau carbon tax, those court fights likely never would have happened.

Gerald Butts, a former senior adviser to Trudeau, spoke to columnist Justin Ling for an article on the carbon tax in the *Toronto Star* in September 2024. He shared an anecdote about Wall's early opposition to the carbon tax.

"That's the base (the national party) will build back from," Butts told his Liberal colleagues at the time. "So listen to what Brad Wall is saying."

On this, Butts was right.

Wall was the ghost of the carbon tax's future.

CHAPTER 6

THE FIRST FATAL FLAW OF THE CARBON TAX

There were signs a few years ago hinting that support for the carbon tax would crumble.

A poll commissioned by the CBC found Canadians' top concern was cost of living, outstripping climate change by thirteen percentage points. That was in 2019 when the federal government first imposed the carbon tax, well before inflation hit a four-decade high in 2022.

"Canadians are deeply concerned about climate change and are willing to make adjustments in their lives to fight it—but for many people, paying as much as even a monthly Netflix subscription in extra taxes is not one of them," read the report from CBC, based on a poll commissioned from Public Square Research and Maru/Blue. "The numbers suggest that while Canadians care about climate change, their financial concerns are more important."

About thirty-two percent of Canadians said they would not be willing to pay *any* tax to prevent climate change. Another seventeen percent said they would only be willing to pay less than $100 in additional taxes per year. To put those results another way: half of Canadians said they would not support a carbon tax that cost them more than $100 annually.

These poll results should have been a wake-up call for the Trudeau government and its plan to impose higher carbon taxes on Canadians year after year. People were struggling, but politicians missed the warning sign.

Worse, politicians skipped the step of seeking consent for their stinging carbon tax plan.

Prime Minister Justin Trudeau swept into office with promises of modest temporary deficits, expanded immigration and legalized marijuana. But voters had to squint to see a hint of the carbon tax policy that would become a pillar of Trudeau's legacy.

Buried thirty-nine pages deep into the eighty-eight-page 2015 Liberal platform was the promise to "put a price on carbon, and reduce carbon pollution."

A few paragraphs later, the Liberals said they would "reduce greenhouse gas emissions" and "end the cycle of federal parties—of all stripes—setting arbitrary targets without a real federal/provincial/territorial plan in place."

The only other mention of a carbon tax came on page forty, with a promise to "establish national emissions reduction targets, and ensure that the provinces and territories have targeted federal funding and the flexibility to design their own policies to meet these commitments, including their own carbon pricing policies."

Canadians could be forgiven for failing to understand just how expensive this commitment would be. In the eighty-eight-page platform, Trudeau's signature carbon-tax policy was mentioned just three times and was unrecognizably vague.

What could Canadians be expected to take away from this?

Not that the government would impose a carbon tax that would make the necessities of life—such as driving to work, heating your home or putting food on your family's table—more expensive. Not that the carbon tax would override provincial jurisdiction over natural resources and the environment, which would trigger constitutional court fights. Not that the vast majority of countries, including most of the world's largest emitters, would refuse to impose carbon taxes.

Not that Canada's carbon tax would have, at best, a negligible impact on global emissions. Not that the Liberals would hike the carbon tax every single year until 2030, and that it would be layered on top of a myriad of other energy taxes and regulations.

All Canadians could meaningfully expect to take away from the Liberal Party's promises in 2015 was the government would be committed to reducing pollution. After all, to any ordinary Canadian, "a price on carbon" and a commitment to "reduce carbon pollution" means little more than a promise to be a respectful steward of the environment. And in 2015, when the economy was good and inflation was low, that hardly merited a second look.

The point of a carbon tax is to reduce emissions by making it more expensive to use carbon intensive fuels. It is designed to make driving vehicles fuelled by gasoline or diesel, or heating homes with natural gas or propane, more expensive. By making these fuels more expensive, carbon tax proponents hope that people will use less of them, or switch to wind or solar energy. That is to say, higher prices are a feature of the carbon tax, not a bug.

A fatal flaw, baked into the carbon tax from the beginning, was that the young Trudeau government was not honest with Canadians about the cost of the carbon tax. And by refusing to be honest with Canadians about what it was proposing—a tax that would increasingly make the necessities of life more expensive—it was also refusing to be honest with itself about support for the carbon tax among the Canadian people.

Eventually, some saw the signs.

"Some in Trudeau's government were convinced they didn't need to listen," wrote *Toronto Star* columnist Justin Ling in 2024. "They were so buoyed by polls showing widespread support for climate action that they were ignoring the more sobering data underneath: Canadians wanted climate action, sure, but they were loath to pay more in taxes to make it happen."

It's tough enough for families to make ends meet, and the carbon tax makes it even tougher. The Trudeau Liberals weren't honest with Canadians about the pain the carbon tax would cause, and they definitely didn't manage to manufacture consent among the people. To make matters worse for the government, this was not the only fatal flaw of the Trudeau carbon tax.

CHAPTER 7

THE SECOND FATAL FLAW OF THE CARBON TAX

The carbon tax's first fatal flaw was that Prime Minister Justin Trudeau never sought or received informed consent from Canadians for his costly scheme. The government sold itself a false bill of goods based on the vague notion Canadians wanted a healthy environment. But that never meant Canadians could afford a carbon tax.

The second fatal flaw emerged when the Trudeau government convinced itself, and tried to convince Canadians, that a carbon tax would make life more affordable.

"The prime minister is going to outline how the system is going to be applied across the country, and how families are going to be better off because of it," a government official told the *Toronto Star*, before the carbon tax was first imposed in 2019.

It's a message Trudeau and his ministers have never missed an opportunity to repeat.

"Eight out of ten families will get more money back than they paid," Environment Minister Steven Guilbeault has often said.

The Trudeau government tried to convince Canadians it could impose a carbon tax, then more than offset the cost by sending people money back through rebates. That was a tough message to sell in 2019 and it only got tougher in the years ahead.

Only fourteen percent of Canadians believe they get more money back in rebates than they pay in carbon taxes, according to a 2023 poll from the Angus Reid Institute.

The carbon tax started at about four cents per litre of gasoline ($20 per tonne of emissions). By 2030, the Trudeau government plans to increase the carbon tax all the way up to thirty-seven cents per litre of gasoline ($170 per ton).

The pain of the annual carbon tax hike wasn't soothed by rebates. Then a moment of political expediency became an implicit admission that the carbon tax stretches family budgets.

Trudeau, surrounded by Atlantic Liberal MPs, held a press conference on "delivering support for Canadians on energy bills" in October 2023. The prime minister cheerily removed the carbon tax from home heating oil for three years. With that decision, Trudeau all but conceded the carbon tax makes life more expensive. If it didn't, why would his government remove the carbon tax from home heating oil to help Canadians with energy affordability?

There were also cold, hard numbers from the federal government's own budget watchdog that rocked Trudeau's affordability angle.

"The average household in each of the backstop provinces will see a net cost, paying more in the federal fuel charge and GST, as well as receiving lower incomes (due to the fuel charge), compared to the Canada Carbon Rebate they receive," reads the latest Parliamentary Budget Officer (PBO) report on the carbon tax.

In fact, the PBO has released *three* separate reports confirming the carbon tax costs average families more money than they get back in rebates.

And the numbers are damning.

The carbon tax will cost the average family up to $399 more this year than they get back in rebates, according to the PBO. By 2030, that cost will jump to $903.

Those are annual costs, which means even those figures downplay the cumulative cost over time. By the end of 2030, the carbon tax will have cost the average household up to $6,550, even after the rebates.

But even that undersells the true cost. In its latest report, the PBO only crunched the cost of Trudeau's *consumer* carbon tax. But Trudeau also imposed a hidden carbon tax on *businesses* and buried a third carbon tax in fuel regulations (more on the other carbon taxes later).

Environment Minister Steven Guilbeault claimed the PBO report showed Canadians are better off with the carbon tax. He called claims to the contrary a "big lie" and "misleading."

But the government's claims are based on simplistic math that focuses on the money collected through the carbon tax and the money paid out through the rebates. In doing so, the government conveniently ignores the overall economic costs of the carbon tax.

The PBO's analysis takes a more comprehensive approach.

Carbon taxes cost so much because they hit almost all aspects of Canadian life.

As of the beginning of 2025, the carbon tax adds about $13 to the cost of fuelling up a minivan, about $20 to the cost of fuelling up a pickup truck and about $200 to the cost of fuelling up a big rig.

Canada is a cold place, and the carbon tax currently adds about $386 to the cost of the average family's annual home heating bill.

Canadians also need to eat, and the carbon tax makes that more expensive too. The carbon tax makes it more expensive for farmers to grow food, and it also makes it more expensive for truckers to deliver food, which in turn makes it more expensive for Canadians to buy food.

The federal government gave farmers an exemption on the carbon tax for diesel and gasoline. That helps farmers keep food prices down and compete globally. But the government failed to provide a carbon tax exemption for the propane and natural gas farmers need to dry their grain and heat their barns. As a result, the carbon tax cost Canadian farmers an average of $14,000 in 2019, according to the Canadian Federation of Independent Business. Trudeau has cranked up his carbon tax every year since, which meant higher costs for farmers and higher grocery prices for Canadians. The carbon tax on propane and natural gas will cost farmers $1 billion through 2030, according to the PBO.

"In 2024, the carbon tax will add just under $2 billion to annual trucking costs in Canada," according to the Canadian Trucking Alliance (CTA). "Over the twelve-year tax phase in, the tax will have cost the trucking industry more than $26 billion."

Those costs are then passed on to consumers, including through higher prices on all the groceries truckers bring to the store.

"Due to razor thin margins in the trucking industry, these added costs cannot be absorbed and must be passed on to customers," the CTA said. "As virtually every good purchased by Canadian families and businesses involves truck transportation, this means those families and businesses are paying increasingly higher prices for those goods to pay for this ineffective tax."

And the federal government *knows* all of this.

Data from Environment and Climate Change Canada found the carbon tax would create a $12 billion hole in the Canadian economy in 2024 *alone*.

If that wasn't bad enough, not only does the federal government carbon-tax your fuel, it also applies its sales tax on top of the carbon tax. So first the government taxes the gas, then it taxes the tax. This carbon tax-on-tax alone cost Canadians $400 million in 2024, and will cost Canadians $4 billion by the end of 2030. And that's money the government doesn't even *pretend* to rebate back to Canadians.

Not only does the carbon tax make nearly all aspects of Canadian life more expensive, it also cost taxpayers hundreds of millions of dollars to hire the bureaucrats needed to administer the carbon tax and rebate scheme. The taxpayer-borne price tag is already $283 million, according to government records. Between 2025 and 2030, it is projected to cost taxpayers another $513 million.

Trudeau and his ministers tried to convince Canadians that somehow they could make nearly everybody better off with the carbon tax. But common sense tells you that a government can't impose a carbon tax on its citizens, charge its sales tax on top of it, spend hundreds of millions of dollars hiring hundreds of bureaucrats to administer it, then magically make everyone better off with rebates.

Nevertheless, the Trudeau Liberals continued to push their affordability talking points, despite the fact they were proven wrong by the PBO—not once, not twice, but three times.

The government stuck to its messaging even when Trudeau's own actions—providing a carbon tax break on home heating oil—shattered the affordability argument. In fact, the government tripled down on its affordability spin despite common sense dictating that Canadians can't give the government $20 and get $50 back.

Canadians saw through the government's claims because they knew a tax on the necessities of life wouldn't make things more affordable. And by leaning so heavily on its erroneous affordability pitch, the Trudeau government left itself extremely vulnerable on its signature policy.

Before first imposing the carbon tax in 2019, the Trudeau government decided to frame the tax and rebate scheme as a way to make life more affordable. They never stopped to check in with already overtaxed Canadians to see if they could afford to pay more taxes. And the polling in 2019—when inflation was just 1.9 percent—showed most were not willing to pay more, let alone what the Trudeau government would demand as inflation reached a four-decade high in the years to come.

While Trudeau and the Liberals were riding a wave of popularity, they overlooked fatal flaws in their signature policy that would inevitably turn into fatal blows.

First, they were never completely transparent with Canadians. When crafting the carbon tax, the government never sought or received informed consent from the people about what it was proposing and what the true costs would be.

Second, they never stopped to ask if Canadians could afford to pay higher taxes.

The carbon tax saga is full of twists and turns, but these two fatal flaws doomed the policy from the start.

CHAPTER 8

UNEQUAL APPLICATION OF THE CARBON TAX

Prime Minister Justin Trudeau's carve-out for home heating oil wasn't the only way the carbon tax had been applied unevenly and unfairly across Canada. It turns out in the eyes of the Trudeau Liberals, all provinces are equal, but some provinces are a little more equal than others.

The federal government was not the first government in Canada to impose a carbon tax. Prior to 2019, provincial governments in British Columbia (2008), Quebec (2012), Ontario and Alberta (2017), and Manitoba (2018) had already introduced carbon taxes of their own.

The federal carbon tax cost about four cents per litre of gasoline, five cents per litre of diesel and four cents per cubic metre of natural gas, when it first went into effect on April 1, 2019.

The federal government called this a "national minimum" tax. That meant provinces with carbon taxes in place could keep them, as long as they met the rising mandatory tax rate set by the federal government. Only provinces without a carbon tax in place, or ones which failed to keep pace with the rising tax rate mandated by the Trudeau Liberals, would have the federal tax imposed on them.

Trudeau made this clear in 2016, when he stood in the House of Commons and outlined his vision for the carbon tax, warning that if a province did not impose one, "the government of Canada will implement a price in that jurisdiction."

But as soon as 2019 rolled around, and the federal carbon tax came into effect, it became clear the policy wasn't being implemented the way Trudeau promised.

Initially, the federal carbon tax was directly applied in New Brunswick, Ontario, Manitoba and Saskatchewan, while the other provincial governments were allowed to impose their own carbon taxes. Inequality quickly became apparent.

Provincial carbon taxes were supposed to be just as costly as the federal tax. In theory, the federal government's mandatory minimum meant "carbon pricing is in place across Canada at a similar level of stringency."

But in practice, the federal government's "national minimum" turned into a patchwork of carbon taxes, where some taxpayers in some provinces, particularly in the West, were forced to pay more, while other taxpayers in other provinces, particularly in the East, were allowed to pay less.

The Canadian Taxpayers Federation (CTF) held press conferences across the country to call attention to Trudeau's unequal treatment of Canadians in May 2019, less than two months after the federal government imposed its carbon tax.

At that time, the mandatory minimum carbon tax rate across the country was supposed to be about four cents per litre of gasoline. But the effective carbon tax rate in Nova Scotia was less than one cent per litre. In Prince Edward Island, the effective rate was one cent per litre. And in Newfoundland and Labrador, it was just 0.42 cents.

The Nova Scotia government even bragged publicly about this unfairness.

"The [carbon tax] will add about one cent per litre to the price of gas, compared with about eleven cents per litre by 2022 under the federal approach," stated Nova Scotia's website.

Meanwhile, P.E.I.'s carbon tax was equivalent to just one cent per litre of gasoline because the province had cut fuel taxes by 3.4 cents per litre to offset the cost.

The government of Newfoundland and Labrador had done the same, going even further by cutting provincial fuel taxes by four cents per litre. The province was also allowed to exempt home heating from the carbon tax.

For years, this patchwork of carbon taxes allowed Atlantic provinces, particularly Nova Scotia, to apply lower carbon tax rates. But the special treatment ended in 2023 and the political situation exploded.

The federal government imposed its carbon tax in Atlantic Canada on July 1, 2023, to bring that patchwork in line with other provinces.

Overnight, gas prices in Nova Scotia jumped by 12 cents per litre—the single largest carbon tax increase in Canadian history. And the huge cost this imposed on Nova Scotia taxpayers led not only to hardship, but swift backlash, which in turn fuelled dissent from the province's political leaders.

As polling for the Liberals in the region slumped, Trudeau exempted the carbon tax from home heating oil in October 2023 in an attempt to stop the bleeding.

Home heating oil is a fuel source used almost exclusively in Atlantic Canada. What began as an attempt to reduce the myriad of carbon taxes in Canada, and thereby even the playing field between provinces, resulted in a new carve-out that further fuelled rising regional tensions.

But Atlantic Canada was not the only place to get a significant break on its carbon tax bills.

Back in 2019, Quebec's carbon tax (officially known as a cap-and-trade system), met the "national minimum." But Trudeau never required Quebec to increase its taxation levels in lockstep with the federal carbon tax, as other jurisdictions were forced to. As a result, taxpayers in Quebec pay a significantly lower carbon tax to this day.

Currently, taxpayers in every province and territory are required by the federal government to pay a carbon tax of $80 per ton. Quebec's carbon tax is set at $49 per ton, based on its November 2024 cap-and-trade auction.

For an apples-to-apples comparison, in November 2024, the carbon tax in the rest of Canada costs seventeen cents per litre of gasoline, but only twelve cents in Quebec. In the rest of Canada, the carbon tax costs twenty-one cents per litre of diesel, but only fifteen cents in Quebec. In the rest of Canada, the carbon tax costs fifteen cents per cubic metre of natural gas, but only nine cents in Quebec.

In 2030, Quebec's carbon tax is forecasted to reach twenty-three cents per litre of gas. By then, all other Canadians will be paying thirty-seven cents per litre.

That means in 2030, an Ontario family will pay $10 more in carbon taxes when they fuel up their minivan than drivers in Quebec. Even *La Presse* notes the Quebec carbon tax "is much less than in the federal system."

The Trudeau government claims "carbon pricing is in place across Canada at a similar level of stringency." But if that was the case, why do Quebecers pay less at the pumps?

Compare the cases of Quebec and Nova Scotia.

Until the summer of 2023, Nova Scotia had a provincial cap-and-trade carbon tax similar to Quebec's. Nova Scotia had reduced its emissions by thirty-five percent since 2005, while having by far the lowest carbon tax rate on gasoline in Canada. Meanwhile, Quebec had reduced its emissions by just eight percent.

And yet, Ottawa let Quebec keep its lower rate, while Nova Scotia was hit the largest carbon tax hike in Canadian history courtesy of the Trudeau Liberals.

Trudeau promised the carbon tax would be applied fairly and evenly across Canada. But the preferential treatment the Trudeau government showed, first to Atlantic Canada, and to Quebec even to this day, hasn't gone unnoticed among politicians in the rest of the country.

"You pay more, Quebec pays less," Alberta Premier Danielle Smith said. "The Liberal-NDP coalition are cutting a 'special deal' for Quebec to pay less carbon tax, while raising the price on other provinces, including our own."

CHAPTER 9

CARBON TAX COURT FIGHTS

The carbon tax was supposed to be a cooperative effort with the provinces. Instead, it almost immediately devolved into protracted court fights that pitted most provinces against Ottawa and split the most distinguished jurists in the nation.

Saskatchewan Premier Scott Moe kicked off the court fight with a constitutional challenge in April 2018, a full year before the carbon tax took effect.

"We do not believe the federal government has the constitutional right to impose the Trudeau carbon tax on Saskatchewan, against the wishes of the government and people of Saskatchewan," Moe said.

The Saskatchewan government argued the constitution gives provincial governments jurisdiction over natural resources and the environment. As a result, the federal government did not have the right to override provincial authority by imposing a carbon tax.

"Simply put, we do not believe the federal government has the right to impose a tax on one province but not others just because they don't like our climate change plan," Saskatchewan Justice Minister Don Morgan said.

The Ontario government then announced a second constitutional challenge against the federal carbon tax in August 2018. That was followed by yet another constitutional challenge in June 2019 by the

Alberta government. The Canadian Taxpayers Federation (CTF) was granted official intervener status in all three cases in order to fight on behalf of the Canadian people.

In the Saskatchewan case, nine intervener groups, including Environment Defence Canada, the David Suzuki Foundation and Climate Justice, supported the federal carbon tax in court.

The CTF and the Agricultural Producers Association of Saskatchewan were the only non-government organizations fighting against the federal carbon tax in the case. The situation was similar in Ontario, where the CTF was one of just two interveners opposed to the tax.

The CTF's lawyers challenged Ottawa's argument that the carbon tax would be effective in the fight to lower global emissions. The CTF also argued that a carbon tax is, in fact, a tax—not a "fee," a "levy" or any other euphemism the federal government preferred to use. Lastly, the CTF challenged the constitutionality of carbon tax legislation because it allows the prime minister to impose carbon taxes in any region and at any level without letting members of Parliament vote. Only Parliament has the right to impose taxes on Canadians, according to the constitution.

Ultimately, the court ruled in favour of the federal carbon tax in both cases, but it was close.

In Saskatchewan, the court ruled three-to-two that the carbon tax was constitutional. The dissenting judges in Saskatchewan determined the carbon tax was indeed a tax, that regulating emissions was provincial jurisdiction and that Parliament cannot intrude upon it. The two dissenting judges went so far as to label Trudeau's carbon tax "constitutionally repugnant."

In Ontario, the court ruled four-to-one in favour of the federal government. But again, the dissenting judge argued the federal carbon tax is an unconstitutional invasion of provincial powers and wondered if it would open the door to Ottawa interfering in other inappropriate ways.

The governments of Saskatchewan and Ontario appealed the rulings to the Supreme Court.

In the meantime, Alberta launched its own constitutional challenge after getting rid of its provincial carbon tax.

The Alberta court ruled four-to-one that the carbon tax was unconstitutional in February 2020. The court found that the carbon tax tramples on provincial jurisdiction—going so far as to call it a "Trojan horse" that "substantially overrides" the constitution. The Alberta ruling built on the dissenting opinions in Saskatchewan and Ontario, adding to the momentum taxpayers enjoyed heading into the Supreme Court case.

Following the Alberta victory, a total of fifteen judges had weighed in on the constitutionality of the Trudeau carbon tax, with eight ruling in favour and seven against.

Before the Supreme Court case, one other interesting and surprising, development occurred: the government of Quebec intervened to oppose the federal carbon tax.

"[The Quebec government] will reiterate the importance of protecting provincial autonomy and provinces' capacity to make their own choices in their areas of jurisdiction in its [carbon tax court] intervention," said Quebec Justice Minister Sonia LeBel.

In 2019, it was hard to imagine anything uniting politicians in Alberta and Quebec. But opposition to federal interference via the carbon tax brought them together.

The Supreme Court carbon tax case took place in September 2020. The CTF was the only non-government intervener sticking up for taxpayers and fighting the federal carbon tax during the case.

In addition to the army of taxpayer-funded federal lawyers, the pro-carbon tax side also had lawyers from Environment Defence Canada, the David Suzuki Foundation, Climate Justice and the Canadian Labour Congress, among other groups. The online news site *Blacklock's Reporter* later revealed the Trudeau government used taxpayer money to bankroll pro-carbon tax interveners at the Supreme Court under the guise of "human rights."

The Supreme Court ruled the carbon tax constitutional by a vote of six-to-three on March 25, 2021. Ottawa won, but it was a problematic victory. First, the Supreme Court merely accepted the federal government's argument that the carbon tax was legal, not that it was good policy or that Canada must have a carbon tax. Second, the Supreme Court's decision left open the potential for future challenges against the carbon tax, which was later taken up by the Alberta government.

Before the 2024 New Brunswick provincial election, premier Blaine Higgs promised to launch a renewed carbon tax court challenge. When the federal government removed the carbon tax on home heating oil, the move disproportionally helped taxpayers in Nova Scotia and Prince Edward Island, but left out those in other provinces, including New Brunswick. Federal carbon tax "carve-outs violate the Supreme Court's ruling," New Brunswick's Progressive Conservative Party said, when announcing Higgs' plan to go back to court.

Higgs lost his re-election bid, but Alberta Premier Danielle Smith picked up the torch. In October 2024, Alberta announced a renewed court challenge against the carbon tax over the Trudeau government's three-year carve-out for home heating oil—a move that was seen as nakedly political and designed to primarily benefit people in Atlantic Canada.

When the Supreme Court originally okayed the federal carbon tax in March 2021, it was because the government argued that climate change was a national problem that required a national solution. The court ruled Ottawa could set minimum national standards for emissions reductions under the constitution's Peace, Order and Good Government clause.

"The proposed matter of establishing minimum national standards of GHG price stringency to reduce GHG emissions is of clear concern to Canada as a whole," the court wrote, while noting "the withdrawal of one province from the scheme would clearly threaten its success."

But Trudeau torpedoed that argument in October 2023 by introducing a carve-out that mostly benefited one part of the country.

"Unlike the rest of Canada, fuel oil makes up a large share of residential heating energy in Nova Scotia at thirty-two percent," according to a report from the government of Nova Scotia. "Only Prince Edward Island depends more heavily on fuel oil for residential heating. Across Canada, fuel oil makes up just three percent of residential heating energy, with very low usage in all provinces from New Brunswick to British Columbia."

In fact, there are so few homes using home heating oil in Alberta, Saskatchewan and Manitoba that data for those provinces was listed as "n/a" in the report.

By providing a break to specific regions, Trudeau undermined the constitutional justification that allowed the carbon tax to exist in the first place.

Only weeks before Trudeau removed the carbon tax from home heating oil, Ken McDonald—a Liberal MP from Newfoundland and Labrador—voted to "repeal all carbon taxes."

"Everywhere I go people come up to me and say, 'You know, we're losing faith in the Liberal Party,'" McDonald told *CBC*. "Seniors who live alone tell me they go around their house in the spring and winter time with a blanket wrapped around them because they can't afford the home heating fuel."

Trudeau then announced his carbon tax carve-out surrounded by Atlantic Liberal MPs, and credited those "amazing" MPs as the reason folks in the region got relief.

Others extended that logic.

"Perhaps they need to elect more Liberals," said Gudie Hutchings, minister of the Atlantic Canada Opportunities Agency, when asked why most Canadians weren't getting relief.

Justice Malcolm Rowe foresaw this issue in his dissent in the Supreme Court decision. He wrote that "regulations that have the effect of favouring or imposing unequal burdens on certain provinces and industries in a manner that cannot be justified" would be unconstitutional.

Rowe's concern is now more relevant than ever. Trudeau's carve-out disrupts the very uniformity that made the carbon tax constitutional. The Trudeau government's decision to selectively enforce the tax has weakened its position. If the carbon tax is no longer a uniform measure, it's no longer a "national" policy.

Higgs was right to suggest taking the carbon tax back to court over Trudeau's carve-out. And Smith was right to pick up the torch and launch a new constitutional court fight to dismantle Ottawa's carbon tax and return power to where it belongs—the provinces.

CHAPTER 10

THE CARBON TAX DOES NOT REDUCE EMISSIONS

From the beginning, the Trudeau government pushed the carbon tax on Canadians as the best way to help the environment and reduce emissions.

"Canada needs to cut greenhouse gas emissions that cause climate change and the best way to do that is to put a price on carbon pollution," said Finance Minister Bill Morneau in 2018, when announcing the incoming carbon tax.

That was far from the only time the Trudeau government made this claim.

"Putting a price on pollution is the most efficient and powerful way to keep 1.5 alive," Prime Minister Justin Trudeau said in 2021, referring to the push to keep global temperature increases to 1.5°C.

The messaging remained consistent for years.

"Putting a price on pollution remains the most effective way to fight climate change," the federal government said in a press release in 2022, as it announced plans to hike the carbon tax every year until 2030.

Even taking these claims at face value, it's clear that a carbon tax in Canada wouldn't be effective at lowering global emissions, absent other countries also imposing carbon taxes of their own—in particular, the largest emitters. That's because Canada only makes up 1.4 percent of global emissions, a fact Trudeau has acknowledged.

"Even if Canada stopped everything tomorrow, and the other countries didn't have any solutions, it wouldn't make a big difference," Trudeau said in 2018 during an appearance on the popular Quebec talk show *Tout Le Monde En Parle*.

The Canadian Climate Institute (CCI) is a leading advocate for carbon taxes and its data show the federal carbon tax doesn't move the needle on global emissions. If Canada removed every single climate policy (not just carbon taxes), our emissions would only be twenty-three percent higher this year, according to the CCI. That means with every last climate policy eliminated, Canada would only emit an extra 145 megatonnes of carbon dioxide this year.

The entire world produced 48,210 megatonnes of carbon dioxide in 2021, according to government data. So even if Canada eliminated every single climate policy, the world's emissions would increase by less than one-third of one percent. In fact, even if Canada's emissions fell to zero, global carbon emissions would only be cut by 1.4 percent.

And Canada's share of global emissions is expected to decline even further in the years to come because of the "rapid increase in emissions from economically developing and emerging countries," according to the government of Canada.

The Parliamentary Budget Officer went so far as to state "Canada's own emissions are not large enough to materially impact climate change."

It's clear that a carbon tax that punishes Canadians for the sins of fuelling up their vehicles and heating their homes won't move the needle on global emissions.

The math is clear: Canada's carbon tax would only work to help reduce global emissions if the government could convince other countries to impose carbon taxes too (and if those taxes actually lowered carbon output).

It's a bit like trying to lift a piano onto a truck. It's easy if you have a bunch of people lifting at the same time. It's impossible if you don't.

And most countries are not lifting with a carbon tax.

About seventy percent of countries don't have a national carbon tax, according to World Bank data. Four of the five largest emitting countries—the US, India, Russia and Brazil—don't impose national carbon taxes. And if China's self-reporting is accurate, its carbon tax is seventy-eight percent lower than Canada's, despite the fact its emissions are eighteen hundred percent higher.

One of the major reasons why Trudeau failed to sell Canadians on the carbon tax being a legitimate environmental solution was because he failed to sell carbon taxes to the international community, despite Trudeau's vows to turn Canada into a global leader on climate action.

"Canada will take on a new leadership role internationally," Trudeau said at the United Nations 2015 Climate Conference. And in 2018, just before imposing his carbon tax on Canadians, Trudeau said: "Climate change is a global challenge that requires a global solution."

The problem for Trudeau is that other countries didn't follow his lead on carbon taxes.

This is best illustrated by our biggest ally and trading partner—and the world's largest economy—the United States.

The US federal government hasn't imposed a carbon tax on Americans. Democratic presidents and candidates ranging from Barack Obama to Hillary Clinton, Joe Biden and Kamala Harris didn't impose or run on carbon taxes. And neither have Republicans.

"Mr. Obama and other top administration officials no longer call publicly for a national price on carbon," the *New York Times* reported in 2014.

"We have done extensive polling on a carbon tax," said John Podesta in 2014, who would serve as Hillary Clinton's presidential campaign chair in the 2016 race. "It all sucks."

"The [climate] community has largely moved into a different framework," Podesta said in 2020, when asked if the Biden administration would impose a carbon tax.

There were no mentions of a carbon tax in Obama's 2012 platform, Biden's 2020 platform or Harris' 2024 platform. Serious Democratic contenders for the White House all backed away from carbon taxes.

President Donald Trump didn't impose a carbon tax during his first term in office and there no was no mention of a carbon tax in his 2024 platform.

But that didn't stop Trudeau from trying to convince the rest of the world.

At the United Nations 2021 Climate Conference (COP26), Trudeau announced his government's latest attempt to push carbon taxes on other countries.

"One of the things we all know needs to come out of COP26 is a clearer call to create a global standard around putting a price on pollution," Trudeau said.

He let the cat out of the bag, admitting that citizens in countries with a carbon tax are being penalized. Global carbon taxes would "ensure that those who are leading on pricing pollution don't get unfairly penalized," Trudeau said.

At the conference, Canada announced the launch of its Global Carbon Pricing Challenge—an attempt to get other countries on board with carbon taxes. The government set a goal of having "sixty percent of global GHG emissions covered by carbon pricing policies by 2030." The program website notes "carbon pricing is most effective when more countries adopt it."

The government is spending $1.7 million on the Global Carbon Pricing Challenge (GCPC). Only a dozen countries have signed onto the GCPC as "partners," alongside the European Union.

Canadians pay high carbon taxes, but citizens in the vast majority of countries do not.

This leads to what is known as carbon leakage. When one country imposes strict carbon taxes and energy regulations, it does not mean global emissions will go down. And that's because strict carbon taxes and energy regulations encourage businesses to leave and set up shop in other, more competitive countries.

So not only does the carbon tax make life in Canada more expensive and make our business environment less competitive, it also incentivizes companies to move production to other countries that have fewer taxes and regulations. And this isn't just a wonky theory. It is easy to see how carbon leakage plays out in everyday life.

To take just one example, Canadians who live close to the US-Canada border often choose to fuel up their cars in border towns in the United States.

"BC drivers facing sky-high gas prices are not only filling their tanks stateside, but also filling containers to bring back with them," reported the *Vancouver Sun* in 2019. "Motorists tired of paying almost $1.70 a litre are flocking to the US to pay as little as $1.13 CDN a litre at Costco. That works out to more than $25 in savings on a fifty-litre fill. One gas station employee said on Wednesday about half the Canadians who come for a top-up also fill extra containers to take home."

Or consider this: when a Canadian family vacations in Kelowna, BC, they a pay carbon tax. But when that same Canadian family goes to Disney World, the vacation is carbon-tax free.

Anyone operating a business that competes with our neighbours south of the border can tell you how hard it is to compete with carbon-tax-free states.

"My competitors to the south of me in the United States do not pay that [carbon] tax, so now my cost goes up and I have no alternative," said Jeff Barlow, a corn, wheat, and soybean farmer in Ontario. "By penalizing me there's nothing else that I can do but just be penalized."

The Fraser Institute has warned about carbon leakage: "These higher costs of doing business will not cause an immediate exodus of industries, but as economists point out, capital is liquid and new investment in activities burdened by the carbon tax could flow out of Canada and take their emissions elsewhere, possibly to jurisdictions with lower environmental standards, reducing any benefits of the carbon tax. Carbon leakage is a bit like a game of Whack-A-Mole. Yes, with carbon taxes in Canada, we may reduce domestic emissions. But in reality, we'll likely whack them down here to see them pop up elsewhere."

In fact, even pro-carbon tax politicians and activists recognize carbon leakage. Former Liberal British Columbia premier Christy Clark referred to the notion in 2016 when she refused to increase her carbon tax.

"What happens in situations like that is polluters just move right across the border and pollute where it's cheap, and we want to make sure we fight pollution across Canada and across the world," Clark said. "We will consider raising the carbon tax once other provinces catch up."

Carbon tax activists also recognized the issue of carbon leakage when they called for a Canada-wide carbon tax.

"A carbon tax should be applied across the entire Canadian economy," reads a report from environmental law charity Ecojustice. "A tax on carbon broadly across the economy ensures that all sectors respond according to the price signal, thus maintaining a level economic playing field for all businesses."

The report also acknowledged that, "If the carbon tax is not economy wide, carbon intensive sectors that have been exempted from the tax will have an advantage over low carbon producers as well as over those carbon intensive competitors who have internalized the cost of GHGs into their products. This would artificially skew the market toward exempted producers at the expense of everyone else. A federal, nationwide tax represents a unified and simplified approach to reducing emissions and avoids the border related headaches of a fragmented province-by-province carbon price."

Let's put that more clearly. The carbon-tax activists were pushing for a national carbon tax. If the carbon tax was not nationwide, then businesses in provinces with no carbon tax would have a competitive advantage over businesses that were forced to pay the carbon tax. This would encourage more businesses to produce in the provinces where there was no carbon tax.

That same rationale extends globally. With most countries not imposing national carbon taxes, businesses paying the Canadian carbon tax are at a disadvantage compared to their competitors abroad. Canada's carbon tax incentivizes businesses to set up or increase production in other countries that do not have carbon taxes.

In fact, Ecojustice acknowledged this problem in their own report.

"A handful of carbon intensive and trade-exposed industries could experience a drop in growth to such a degree that they may seek to shift production to other countries with a lower carbon price," according to Ecojustice. "Protecting Canada's economy from other countries with lax or non-existent carbon pricing is important."

This obvious fact—that businesses will go to jurisdictions with lower carbon taxes and regulations—is why carbon tax activists push for global carbon taxes.

"One of the primary challenges of the fragmented carbon pricing system is carbon leakage," notes the *Leaf*, an online publication focused on sustainability and climate news. "This relocation of production does not reduce global emissions but simply relocates them, thereby undermining the environmental objectives of carbon pricing."

Here's another key point: even if other countries imposed national carbon taxes in lockstep with Canada, it doesn't necessarily mean emissions would go down. And that's because a carbon tax is a tax on the necessities of life. Fuelling up a car with gas, heating homes and businesses with natural gas, drying grain with propane, or filling up a big rig with diesel, none are optional.

A carbon tax doesn't stop people from engaging in these activities, it just makes them pay more for the privilege. And that means they're forced to cut back elsewhere in their budgets, like skipping an annual family vacation or socking away less money for their children's university educations. Instead of reducing emissions, carbon taxes reduce the amount of money people have to spend on other areas of their lives.

To illustrate this point, let's take a closer look at British Columbia's experiment with carbon taxes. It is an illuminating case study for all Canadians because the Trudeau carbon tax was, in many ways, modelled on BC's carbon tax.

BC first imposed its carbon tax in 2008. The provincial government claimed it would reduce emissions by a third in twelve years.

"By 2020 and for each subsequent calendar year, BC greenhouse gas emissions will be at least thirty-three percent less than the level of those emissions in 2007," reads the Greenhouse Gas Reduction Targets Act.

Politicians also repeated this claim directly to British Columbians.

"We said in British Columbia it's going to be thirty-three percent," BC premier Gordon Campbell said in 2009. "We all support that. That's going to require … a carbon tax."

The BC government publishes its emissions data annually.

In 2007, the year before BC's carbon tax was implemented, the province emitted 65.5 megatonnes of carbon dioxide equivalent, according to the data. At the beginning of 2020, BC's emissions reached 66.1 megatonnes. Instead of declining by thirty-three percent, BC's emissions increased.

During the pandemic years of 2020 and 2021, when the economy was at a standstill due to government-imposed lockdowns, emissions temporarily dipped. But as of 2022, which is the latest year for which government data has been published, BC's emissions are again higher than they were before its carbon tax was imposed.

The takeaway from the BC experience is clear: politicians promised the carbon tax would reduce emissions, but emissions went up. And at the federal level, the government doesn't even know how much the carbon tax is reducing emissions, if at all.

Conservative MP Dan Mazier asked if the federal government measures "the annual amount of emissions that are directly reduced from federal carbon pricing, and, if so … what is the amount of emissions that have been reduced in Canada directly and specifically from federal carbon pricing, broken down by year?"

Here's the response from Environment Minister Steven Guilbeault, on January 29, 2024: "The government does not measure the annual amount of emissions that are directly reduced by federal carbon pricing."

The Trudeau government doesn't know how much the carbon tax is reducing emissions, because the Trudeau government is not tracking this information for itself.

The government's target is to reduce emissions by forty to forty-five percent below 2005 levels by 2030. But the government looks unlikely to hit that target, according to a 2024 report from the commissioner of the environment. As of 2022, Canada's emissions have fallen by a mere 7.1 percent below 2005 levels.

"There's only six years left to do essentially twenty or thirty years' worth of reductions," Environment Commissioner Jerry DeMarco said in November 2024.

Trudeau tried to sell carbon taxes to Canadians as an environmental policy, but it could never work when the vast majority of countries, including most of the world's largest emitters, don't have national carbon taxes.

The ultimate fatal flaw of the carbon tax is that it doesn't work, and this fatal flaw was baked into Trudeau's carbon tax cake from the beginning.

In fact, others are doing better emissions-wise than Canada without a carbon tax.

Take a look at our neighbour to the south. Canada has a carbon tax, the United States doesn't. "But as of 2021, emissions had fallen by only

8.5 percent in Canada and 15.2 percent in the US," according to a 2024 report from TD Economics.

A 2024 poll released by the Angus Reid Institute makes clear that Canadians don't believe Trudeau when he says the carbon tax is working. According to the poll, sixty-eight percent of Canadians don't think the carbon tax is effective at reducing emissions.

The Trudeau government tried to sell Canadians on the hope a federal carbon tax in Canada was the "best way" to cut emissions. But the carbon tax is not an environmental solution.

A carbon tax in Canada won't reduce emissions in the US, China or anywhere else. And without international cooperation, a carbon tax in Canada is a drop in the ocean. BC had a carbon tax in place for more than a decade and emissions went up. And the federal government doesn't know if, or how much, its carbon tax is reducing emissions in Canada.

CHAPTER 11

WHY CARBON TAXES DON'T WORK

A carbon tax is meant to drive up the price of fuel in order to change behaviour. Carbon tax advocates claim this is more "efficient" because it makes those who buy fuel pay more to compensate for the costs they put onto society by emitting carbon. With the higher cost of fuel, people will use less fuel, or will be incentivized to choose forms of fuel that are less damaging to the environment. That's the theory.

William Nordhaus, an economist who won the Nobel Prize for his work on carbon taxes and climate change economics in 2018, lays out the reasoning behind carbon taxes:

> First, it will provide signals to consumers about which goods and services are carbon intensive and should therefore be used more sparingly. Second, it will provide signals to producers about which inputs are carbon intensive (such as coal and oil) and which are low-carbon (such as natural gas or wind power), thereby inducing firms to move to low-carbon technologies. Third, it will give market incentives for inventors, innovators, and investment bankers to invent, fund, develop, and commercialize new low-carbon products and processes.

And that's the rationale that led the Trudeau Liberals to make your gasoline and home heating more expensive. Increasing the price of gas, diesel and propane would mean fewer people buy it, which in turn would result in less carbon dioxide emitted. Voila, problem solved.

But it isn't that simple.

The carbon tax rests on some shaky assumptions that complicate matters in the real world, and render its implementation outside an academic paper more difficult. As a result, taxpayers end up with a lot of carbon tax pain for very little—if any—environmental gain.

Even to its proponents, carbon taxes can only be efficient if the government removes all other types of environmental taxes and regulations. This is because "regulations that phase out coal, cap methane emissions, and mandate fuel standards are unnecessary if there is already a carbon price in place signalling the cost of emissions to individuals and firms," as outlined in a 2020 paper from the Fraser Institute.

Economists call this approach, which layers carbon taxes on top of regulations, "belt and suspenders," because it uses two things to do one job. But unlike literal belts and suspenders, layering carbon taxes on top of regulations can mean worse results and more taxpayer money wasted. Arik Levinson, the economist who coined the term "belt and suspenders," says with this approach, "regulatory standards will likely reduce the cost-effectiveness benefits of [carbon taxes]."

Even the carbon-tax stalwart Nordhaus says that once the carbon tax has been set "there is no reason to provide extraordinary support for green innovations."

But when Trudeau introduced his carbon tax, he did not then reduce regulations. On the contrary, a quick scan of the government's website shows Ottawa is adding, not reducing, environmental regulations.

There's a whole list. There is the carbon tax, the hidden carbon tax introduced through fuel regulations, the electric vehicle mandate, carbon capture tax credits, restrictions on fertilizer use in agriculture, methane reduction targets and an emissions cap in the oil and gas industry, as well as new emissions limits for the electricity sector, new building and motor vehicle energy efficiency mandates, and many more. The Trudeau government also passed legislation blocking new pipelines.

All these policies together will cost each Canadian worker $6,700, according to the Fraser Institute. By 2030, because of these policies, Canada will have lost 164,000 jobs.

All these regulations and carbon taxes create even more hoops for businesses to jump through. They create "a tangled web of carbon regulations so complex that investors can no longer assess risk and return in our energy systems," according to economist Peter Tertzakian. "When investors can't understand business risk, they take their money elsewhere."

And it's been this way since Trudeau was elected. Since 2015, Canada has seen nearly $670 billion in natural resources projects suspended or cancelled. Leaving these regulations in place while also charging a carbon tax means even more damage to our economy, because of the combination of distortions to market forces.

But the government isn't just diluting its carbon tax with regulations, it's also doing it with other carbon taxes.

In 2019, Trudeau brought in his first carbon tax, then, in 2023, he brought in his second, hidden carbon tax that he buried in fuel regulations. Those regulations require producers to reduce the carbon content of their fuels. If companies can't meet the requirements, they are forced to buy credits. And the cost of those credits are passed onto consumers through higher pump prices.

More recently, Trudeau announced his third carbon tax through a new cap-and-trade system designed to cap emissions from the oil and gas industry. As a result of the emission cap alone, Canada could lose 110,000 jobs by 2040, according to projections from Deloitte. And by 2030, Canada could lose $7.5 billion in investment, according to estimates from the Conference Board of Canada.

The Trudeau government's botched implementation of the carbon tax has hurt Canadians. But even if the government had avoided many of these implementation errors, the economics of carbon taxes are still flawed.

To charge an efficient carbon tax, the government must be able to set it at the right rate, i.e., a tax rate that lowers carbon emissions enough to reduce the effects of climate change, but not one so high that it makes life unaffordable for families. Nordhaus recognizes this when he says that setting a carbon tax too high is like "burning down the village to save it."

Not only is the tax supposed to reduce the amount of carbon emitted, but if calculated accurately, it's also supposed to correct the externality and make the economy more efficient.

But how does the government know what tax rate to charge? The only honest answer is that it doesn't, because it's impossible to calculate. This is because the government lacks the real-world information needed to figure out the correct carbon tax rate. And if any economist was being honest with you, they would acknowledge that value and costs are subjective to an individual. It's impossible to pinpoint a subjective value or cost to an individual, let alone across an entire country or the world.

Carbon tax advocates and academics call this mystery number the "social cost of carbon," but they don't agree on what it is. As noted by energy economist Robert Murphy in his testimony for the United States Senate Committee on Environment and Public Works, "the estimation of the [social cost of carbon] relies on computer simulations of the economy and climate system for hundreds of years into the future, and furthermore depends on many subjective modelling assumptions ... These assumptions can have an enormous impact on the final number, meaning that an analyst can generate just about any SCC he or she wishes by adjusting certain parameters."

When making these calculations, carbon tax advocates take what is observed in the economy and compare it to a hypothetical scenario. Then they conclude the real-world outcome, based on the actions of real people, isn't as good as their projected world. By being able to pick and choose what to include in their calculations, carbon tax advocates place more value on their imaginations than all the people buying and selling goods in the real world. Their solution is to tax you until the real world looks like their imaginations.

Another problem for carbon tax advocates is that most of the costs associated with the social cost of carbon will occur in the future. Carbon tax advocates must therefore take two additional leaps of faith when setting the so-called optimal carbon tax rate.

First, advocates must be able to sum up the net impact of current emissions on the future. At best, this an estimate. As illustrated above, with the infinite potential external impacts of carbon emissions, advocates must make value judgements on what to include and what to keep out of their analysis. Second, advocates must also know, somehow, what individuals in the future, who will be impacted by today's carbon tax, will value—such as, how much economic decline are people in the future willing to exchange for reduced emissions? To answer this question would require a time machine.

Here's the kicker: even if the government was able to set the optimal carbon tax rate, there is still no guarantee it would reduce emissions. That's because of the problem of inelastic demand. Carbon tax theory assumes the higher price of fuel will mean people consume less of it. But that only works if there is a substitute. For many Canadians, there isn't a choice involved when it comes to filling up a car with gas to get to work or using natural gas to heat their home in the winter. They can't make a different choice, so they just end up paying more thanks to the carbon tax.

The Trudeau government gave itself an impossible task in setting the carbon tax rate. It had to be high enough to reduce emissions (it didn't), but it couldn't be too high or it would make life unaffordable for families (it did).

CHAPTER 12

INFLATION, PANDEMIC AND TAX FATIGUE

The carbon tax forces people to pay more for the necessities of life. But the backlash against the carbon tax is about much more than just the price at gas stations or on natural gas bills. And that's because the carbon tax came on top of so many other costs crushing Canadians.

Canadians were already overtaxed by all levels of government before Prime Minister Justin Trudeau's national carbon tax was imposed in 2019. And a forty-year high inflation and a pandemic that hurt private-sector workers and small businesses exacerbated growing anger toward governments that hike our taxes and waste our money.

And while other governments provided tax relief during the pandemic, the Trudeau Liberals repeatedly hiked the carbon tax. Instead of showing Canadians he understood the hardship they were going through and making life more affordable, Trudeau doubled down with carbon tax hike after carbon tax hike.

Even setting aside the carbon tax, Canadians are overtaxed. This is best illustrated through the Fraser Institute's Canadian Consumer Tax Index.

"The average Canadian family now spends more of its income on taxes (43 percent) than it does on basic necessities such as food, shelter, and clothing combined (35.6 percent)," according to the Fraser Institute. "By comparison, 33.5 percent of the average family's income went to pay taxes in 1961 while 56.5 percent went to basic necessities."

Canadians are right to feel like the cost of everything has increased significantly. But taxes have gone up even faster than prices. The total tax bill for the average Canadian family has increased by 2,705 percent since 1961. Meanwhile, the average price Canadians pay for food, shelter and clothing increased by 901 percent during the same time period.

Canadians pay federal and provincial income taxes, federal and provincial sales taxes, federal and provincial fuel taxes, payroll taxes, property taxes, business taxes, capital gains taxes, digital services and online streaming taxes, alcohol taxes, natural resources taxes and import taxes, among others. You name it, they tax it. And now overtaxed Canadians pay a carbon tax to boot.

That staggering tax burden was being borne by Canadians when the pandemic hit and governments started putting the country in revolving lockdowns that lasted the better part of three years. The growing divide between Canadians outside of government and those shielded by the government's golden gates grew so large it was impossible to ignore.

Thousands of workers in the private sector were sent to the ranks of the unemployed, forced to live on government subsidies they would have to pay back directly or indirectly through higher taxes in the future. Businesses, especially small businesses, were forced to shut down, many permanently. Meanwhile, for the vast majority of bureaucrats and politicians within government, their financial fortunes had never been so good.

Since the beginning of 2020, all MPs have taken five pay raises. Even a pandemic didn't stop them from padding their pockets. These pay raises ranged from an extra $24,200 for a backbencher to an extra $48,400 for the prime minister. As of the beginning of 2025, a backbench MP's salary is now $203,100, the salary for ministers is $299,900, and the prime minister rakes in $406,200.

Adding insult to injury, the carbon tax is hiked every year on the very same day MPs give themselves a pay raise: April 1. Like a bad

April Fools' joke, politicians in Ottawa take more money out of taxpayers' pockets the same day they pad their own with higher pay.

And it's not just politicians who have financially benefited since the beginning of the pandemic. So, too, have government bureaucrats.

During the heart of the pandemic in 2020, the government dished out 373,134 pay raises to bureaucrats, followed by 266,646 in 2021, 162,263 in 2022 and 318,067 in 2023. That means over four years, the federal government issued more than one million pay raises to bureaucrats, while their neighbours in the private sector lost their jobs and businesses and struggled to afford rising rents and grocery bills.

On top of the raises came the bonuses. In the pandemic years of 2020 and 2021, the government rubber-stamped $362 million in taxpayer-funded bonuses. About ninety percent of government executives took a bonus, for an average of about $18,000 each. Meanwhile, government departments failed to meet half of their performance targets in those years.

The size and cost of the federal bureaucracy increased rapidly throughout Trudeau's tenure as prime minister, including during the pandemic. Since 2015, the federal bureaucracy has mushroomed by more than 108,000 employees. The cost of the bureaucracy has increased by seventy-three percent under Trudeau.

All the while, the government made headlines for wasteful spending. Here are a few examples:

Trudeau billed taxpayers for a $6,000-per-night luxury River Suite at the Corinthia Hotel during Queen Elizabeth's funeral in London, England. The Corinthia is described as "one of the top luxury hotels in central London," and the $6,000-per-night suite boasts a view of the River Thames, a marble bathroom and "complimentary butler service."

The governor general's week-long trip to Dubai for Expo 2020 cost taxpayers more than $1 million, and somehow her thirty-person entourage managed to rack up an airplane food bill totalling $100,000. They dined on beef Wellington with red jus, buttery chicken tikka masala, apple-and-cranberry-stuffed pork tenderloin and pan-fried chicken scallopini in wine reduction sauce—not exactly like the "airline meals" us normal folk are used to, as Governor General Mary Simon claimed.

Government bureaucrats at one department—Global Affairs Canada—bill taxpayers $51,000 a month on booze, while government employees across the board bill taxpayers $42,000 a month for their tickets to circuses and concerts, balls and ballets, galas and award shows, football and hockey games, dance festivals and musicals.

In effect, the pandemic highlighted a tale of two Canadas: financial pain for workers and small businesses in the private sector, economic gain for politicians and bureaucrats.

The cherry on top was inflation. In a normal year, prices increase by about two percent. The inflation crisis hit in 2021 as annual price increases began to soar above three percent. In 2022, "the annual growth in the Consumer Price Index rose to a forty-year high of 6.8 percent," according to Statistics Canada.

For many Canadians, this was a breaking point. And perhaps there is no better illustration of this harsh reality than food bank usage.

"In 2020, recorded visits climbed to 1.5 million, followed by 2.12 million visits in 2021," reported *Global News*. "In 2022, 2.65 million people visited food banks, marking a fifty-three percent increase. Finally, 2023 saw a record 3.49 million visits to food banks across Toronto."

And consumer prices have continued to rise. Overall, prices have increased by eighteen percent since the beginning of the pandemic.

"The average Alberta household is currently facing financial pressure equal to adding $400 in new expenses each month," University of Calgary economics professor Trevor Tombe wrote in June 2022, the same month inflation reached a forty-year high. "For perspective, the average monthly grocery bill is roughly $650. These higher prices mean households must change what they buy, use savings to cover the extra costs, or accumulate more debt. Each month that goes by these costs continue to grow."

Tombe updated his analysis in May 2024.

"Since February 2020, rent is up twenty percent, food is up twenty-three percent, gasoline is up thirty percent, and mortgage interest costs are up forty-three percent," Tombe wrote. "Average hourly pay, meanwhile, is up only about fifteen percent. The purchasing power of wages in terms of food is therefore down by about seven percent. In terms of gasoline, it's down twelve percent. And in terms of mortgage interest costs, it's down twenty percent."

What did Trudeau and the Liberal government in Ottawa do during the inflation crisis?

They poured gasoline on the fire with yearly carbon tax hikes. Since the beginning of the pandemic and inflation crisis, the carbon tax has increased five times, totalling three hundred percent. In fact, the feds continued to hike the carbon tax even as provincial governments of all political stripes, as well as governments in other countries, cut fuel taxes.

Manitoba's New Democrats suspended the fourteen-cents-per-litre provincial fuel tax for one year. The Liberals in Newfoundland and Labrador cut the provincial fuel tax by eight cents per litre for nearly three years. Ontario's Progressive Conservative government cut its fuel tax by six cents per litre. Alberta's Conservative government suspended its fuel tax for more than a year and a half. Even British Columbia's NDP government paused its planned carbon tax hikes during the pandemic.

Around the world, many countries cut fuel taxes to ease the pain of the pandemic and high inflation. The United Kingdom announced billions of dollars in fuel tax relief. South Korea cut its gas tax by thirty percent. Germany cut its fuel tax by more than thirty cents per litre for three months. The Netherlands cut its gas tax by seventeen cents per litre.

"I've now used and put in place all of the tax measures that I can," said Paschal Donohoe, Ireland's finance minister, after cutting the excise fuel tax and the sales tax on electricity.

India cut its gas tax to "keep inflation low, thus helping the poor and middle classes." Sweden, Australia, Norway, Israel, Italy, New Zealand, Portugal, among other countries and state governments, also cut fuel taxes in recent years.

Meanwhile, a quick trip to the gas station or a glance at their heating bills were all Canadians needed to know Trudeau wasn't easing the burden. While other governments were providing relief, Trudeau was busy cranking up his carbon tax. Even if the Trudeau Liberals were never going to scrap the carbon tax completely, the very least they could have done was make life more affordable for Canadians amid a cost-of-living crunch with some sort of tax relief.

"We know Canadians are facing challenging times right now, people are squeezed between the cost of groceries, rents," Trudeau said during a cabinet retreat in Montreal aimed at "bringing down the cost of living."

Trudeau had options.

For example, the Trudeau government could have ended its pernicious practice of charging tax-on-tax. The federal government charges its sales tax on top of the carbon tax, meaning the sales tax is applied *after* all the per-litre taxes are added. Ending the carbon tax-on-tax alone would save Canadians about $500 million a year. That was a simple way to save Canadians money during the pandemic when fuelling up their vehicles or heating their homes.

The Trudeau government could have taken the carbon tax off all home heating, like it did with its carve-out for home heating oil, which is primarily used in Atlantic Canada.

About ninety-seven percent of Canadian families use other forms of energy to heat their homes. The government could have extended the same relief to all Canadians by taking the carbon tax off all forms of home heating. That would have saved the average family using natural gas about $1,100 over three years.

The Trudeau government could have given farmers relief by making sure Bill C-234 became law, which would remove the carbon tax from the natural gas and propane used on farms. The carbon tax on the natural gas and propane used to heat barns and dry grain will cost farmers $1 billion by 2030, according to the Parliamentary Budget Officer. And passing Bill C-234 wouldn't have just helped farmers, it also would have helped to lower grocery prices for all Canadians.

For years, the rising cost of living had been the top concern among Canadians. For years, the Trudeau government had made life more expensive by hiking its carbon tax. And instead of providing Canadians with some much-needed relief during a pandemic and a cost-of-living crisis, Trudeau kept inflicting more pain with carbon tax hikes.

The carbon tax on its own imposes large costs on Canadians by making gasoline, diesel, natural gas and other fuel sources more expensive. In turn, that makes life in Canada more expensive, whether it's driving to work, heating your home or buying groceries.

But the carbon tax was only the straw that broke Canada's back.

The pandemic and government-imposed lockdowns hurt Canadians in the private sector in a way that politicians and bureaucrats simply didn't experience. When politicians and bureaucrats took raises and bonuses, frustration grew.

Then came the wave of inflation that sent price increases to a forty-year high. Provincial governments and other countries provided their citizens with relief by cutting or suspending fuel taxes, while Trudeau doubled down and made a bad situation worse.

CHAPTER 13

LIBERALS LOSE ON AFFORDABILITY

Ironically, affordability was the capstone of the Prime Minister Justin Trudeau's carbon tax. The government tried to convince Canadians it could impose a tax and still give most people back more money than the carbon tax cost them. It was a tough sell from the start. And ultimately, that capstone fell and collapsed the carbon tax completely.

The affordability angle is critical because it overshadowed actual emissions. The Trudeau Liberals began by telling Canadians the carbon tax was the best way to help the environment and fight climate change. But they didn't stop there. They also claimed the rebates would leave more money in people's pockets. In other words, Canadians could have their cake and eat it too.

As the years stretched on, the Trudeau government's messaging increasingly shifted, focusing more and more on the supposed affordability benefits of the carbon tax. And initially, there were a few factors at play that helped the government push the affordability angle.

When the carbon tax was implemented in 2019, the rate was set relatively low, at four cents per litre of gasoline. At that time, the Trudeau government falsely claimed the carbon tax would increase to eleven cents per litre of gasoline in 2022, with no further hikes beyond that.

"The commitment was to go up to 2022," said environment minister Catherine McKenna during the 2019 election. "There was no intention to go up beyond that, there's no secret agenda. Any decision to move up would be (in) consultation with provinces and territories."

But just a year later, the Trudeau government announced its plan to keep hiking the carbon tax every year until it reached thirty-seven cents per litre of gasoline in 2030. And to this day, the government refuses to rule out future carbon tax hikes beyond that.

When the carbon tax was implemented, inflation was relatively low, hovering around two percent. There was no cross-country recession, no pandemic and no inflation crisis. Given a favourable economic backdrop, the Trudeau Liberals had an easier time selling the carbon tax and its affordability angle. But as the economic climate deteriorated, the carbon tax kept going up and the financial strain hit Canadians harder.

The first Parliamentary Budget Officer (PBO) report about the carbon tax seemed to save the government's affordability narrative. That first report seemed to confirm the government's favourite talking point that the carbon tax rebates leave most Canadians with more money in their pockets.

"Most households will receive higher transfers than amounts paid in fuel charges," according to the PBO's 2019 report. "They will therefore be better off on a net basis because the rebate exceeds the average household carbon cost."

The Trudeau government shouted this finding from the rooftops and never missed an opportunity to waive the PBO report around in public.

"A price on pollution works, it protects the environment for our kids and grandkids, and puts money back into the pockets of Canadians," McKenna said, following the release of the PBO report.

But the PBO turned out to be a two-edged sword and the second edge was much sharper.

When the PBO later crunched the numbers, the analysis included the broader costs of the carbon tax on the Canadian economy. And once it did that, the PBO concluded the carbon tax cost Canadians *more* money than they got back in rebates. In fact, three separate PBO reports would confirm this central fact about the carbon tax.

"The average household in each of the backstop provinces will see a net cost, paying more in the federal fuel charge and GST, as well as receiving lower incomes (due to the fuel charge), compared to the Canada Carbon Rebate they receive," reads the latest PBO report, published in October 2024.

The carbon tax will cost the average family up to $399 more than the rebates they receive in the 2024–25 fiscal year, according to the PBO. By 2030, that cost will jump to $903. And those are annual costs, so even those numbers downplay the total cost over time. By the end of 2030, the carbon tax will have cost the average household up to $6,550, even after the rebates.

That's because the cost of the carbon tax trickles through the entire economy.

Canadians obviously pay the carbon tax on their heating bills and whenever they go to the gas pumps. But there are also hidden costs. Canadians pay the carbon tax at nearly every level of the economic supply chain, with costs also passed on to consumers whenever a product is shipped or delivered by truck. As the PBO's updated analysis shows, these hidden costs, combined with the direct costs of the carbon tax, leave Canadians worse off financially.

And it isn't just the PBO pointing out that the carbon tax hurts Canada's economy. Even government departments have acknowledged this obvious reality.

"Nova Scotians saw prices at the pump increase by fourteen percent in July compared with June," Statistics Canada reported in August 2023, a month after the federal carbon tax was imposed in Nova Scotia. "The introduction of the federal carbon levy in the province and higher wholesale prices contributed to higher gasoline prices."

The Canada Revenue Agency shows the carbon tax increases the price of gasoline by seventeen cents per litre, the price of diesel and home heating oil by twenty-one cents per litre and the price of natural gas by fifteen per cubic metre in 2025.

Or take the Bank of Canada: "If the charge were to be removed from the three main fuel components of the consumer price index (gasoline, natural gas and fuel oil) it would reduce the inflation rate by 0.4 percentage points."

Translation: If the government scrapped the carbon tax, life would be more affordable.

In fact, the government's own data shows the carbon tax will cost the Canadian economy $12 billion in 2024 alone, and by 2030 it will have blown a $30 billion hole in Canada's gross domestic product.

As more and more data revealed the obvious truth that carbon taxes make life more expensive, the Canadian Taxpayers Federation (CTF) amplified that message for Canadians. Every time a PBO report was released showing the true cost of the carbon tax, the CTF issued press releases, wrote newspaper columns, did interviews with journalists and appeared on TV and radio shows. The CTF also bypassed media entirely with social media campaigns, taking the message straight to Canadians on their phones, tablets and computers.

To further undermine Trudeau's claim that the carbon tax was "revenue neutral," the CTF highlighted the fact that the government doesn't rebate the money generated from the GST applied to the carbon tax. This carbon tax-on-tax will cost Canadians $500 million in 2025. And when the Trudeau government issued rebates to Canadians—which it used as an opportunity to herald its affordability message—the CTF would issue press releases highlighting the PBO's findings.

The carbon tax and rebates are "supporting the everyday affordability challenges of Canadians," Environment Minister Steven Guilbeault claimed in January 2024.

"Canadians do need relief from this inflation, but the government is pulling a fast one on Canadians," the CTF said on CTV.

"The Parliamentary Budget Officer's own numbers show that the carbon tax is going to cost more than the rebates."

Economic analysis wasn't the only source of bad numbers for carbon tax proponents—the polling had also taken a turn.

About sixty-two percent of Canadians told the polling firm Leger in August 2024 that the government should scrap the tax-on-tax it layered on the carbon tax. The CTF released the poll at a press conference in front of Parliament Hill. Only twenty-two percent supported the carbon tax-on-tax, while the rest of respondents were unsure. Among those decided on the issue, seventy-four percent of Canadians supported ending the carbon tax-on-tax.

This was the crucial point the CTF made at that press conference: how can the government make most families better off with its carbon tax and rebate scheme if it's skimming hundreds of millions of dollars off the top with its carbon tax-on-tax?

Then the CTF attacked the cost of the tax collectors administering the carbon tax. The CTF's investigative journalist Ryan Thorpe dug up government records revealing the bureaucratic cost of the carbon tax. Between 2019 and 2024, administering the carbon tax and rebate scheme cost taxpayers $283 million, with $800 million in projected costs between 2025 and 2030.

Increasingly, Canadians didn't buy the idea that the carbon tax made life more affordable.

Polling by Angus Reid published in March 2024 revealed that, by a margin of more than two-to-one, Canadians didn't buy the government's affordability spin. Among respondents who said they received rebates, forty-five percent believed they paid more in carbon taxes than they got back. Meanwhile, only nineteen percent of respondents who said they got rebates believed they got more back than they paid.

Think about that for a second. The Trudeau government never passed up an opportunity to claim that "eight out of ten Canadians get more back than they pay." And yet, just nineteen percent of Canadians who said they received the rebates actually believed what the government was telling them about its signature policy.

The government realized it had a problem. But it misdiagnosed the issue. It ignored the strain the carbon tax put on families struggling with affordability. Instead, the government focused on communications.

The government believed the problem with the carbon tax wasn't that it was bad policy, but that Canadians just didn't understand its benefits. In other words, they came to believe the central problem with the carbon tax was that they had failed to do a good job explaining it to the public.

The spin machine went into overdrive.

"The federal government is considering a rebrand of the rebate program for its carbon pricing system in an attempt to tackle what it calls confusion and misconceptions about the scheme," the *Toronto Star* reported in January 2024.

But this notion rested on a shaky assumption. Was the problem *really* that Trudeau and his ministers hadn't talked enough about the supposed benefits of the rebates? The Trudeau government had never passed up an opportunity to repeat that claim. And who gets more media air time than the prime minister, his ministers and their army of communications staff?

But the government bet it all on a rebranding campaign in February 2024.

Previously, when the rebates showed up in the bank accounts of Canadians, they were called a "Climate Action Incentive Payment," or were labelled under vague names like "CANADA FED." Following the rebrand, those cheques were called a "Canada Carbon Rebate."

Soon, government commercials during every hockey game were heralding the rebates. The government budgeted $7 million for the ads in December 2024.

The CTF hit back against the Trudeau government with an anti-carbon tax ad campaign on social media, with two of the videos receiving more than 700,000 views on YouTube.

The rebranding and ad campaign failed and the carbon tax's popularity continued to plummet.

The Trudeau government thought the problem was that Canadians just didn't understand how great the carbon tax was. They thought that if they explained things a little slower to the masses, Canadians would line up in support of carbon taxes. But the problem wasn't a lack of good communication from politicians and taxpayer-funded spinners. The problem was that Canadians understood all too well an obvious reality about the carbon tax.

It makes the necessities of life, like getting around, staying warm and buying food, more expensive. The government can't carbon tax its citizens, charge its sales tax on top of the carbon tax, spend hundreds of millions of dollars administering its carbon tax, then somehow make the majority of Canadians better off with carbon tax rebates.

And if Trudeau's first carbon tax wasn't bad enough, there was more bad news.

Trudeau buried a second, hidden carbon tax in fuel regulations that took effect on July 1, 2023. The regulations require producers to reduce the carbon content of their fuels. If they can't meet the requirements, they must purchase credits that increase costs that are then passed on to Canadians at the pumps. This second carbon tax is layered on top of Trudeau's first carbon tax and it doesn't come with rebates.

In 2030, when Trudeau's second carbon tax is fully implemented, it will increase the price of gas by up to seventeen cents per litre, costing the average family up to $1,157. That year, Trudeau's two carbon taxes combined will increase the price of gas by about fifty-four cents per litre.

To make matters worse, the government's own analysis shows the second, hidden carbon tax will "disproportionately impact lower and middle-income households, as well as households currently experiencing energy poverty," and will especially harm "single mothers" and "seniors living on fixed incomes."

And just like with Trudeau's first carbon tax, his second carbon tax also means significant costs for the Canadian economy. In 2030, the second carbon tax "will result in an overall gross domestic product decrease of up to $9 billion," according to government documents.

Taxpayers will also need to fork over an extra $90 million to fuel the bureaucracy administering this regulatory quagmire. That means Canadians will have the pleasure of paying higher taxes so more federal tax collectors can increase their fuel prices.

This is the simple reality the Trudeau government wasn't able to contend with: carbon taxes make life more expensive.

The cost isn't a bug of carbon taxes, it's a feature. After all, if carbon taxes didn't make life more expensive, then they wouldn't push people to make different choices. That's the reality. And despite its best efforts, the Trudeau government has never been able to get Canadians to ignore reality.

The Trudeau government then ran up against an inflation crisis, which helped to spotlight the way carbon taxes were making life more expensive for Canadians. The Trudeau government also ran into the CTF and our hundreds of thousands of supporters across the country, who constantly reminded Canadians that carbon taxes make the necessities of life more expensive.

It doesn't matter how many government spinners you employ, how many tax dollars you waste on carbon tax ad campaigns, or how many times you rebrand the rebate cheques, reality wins out in the end. And it was the reality that carbon taxes make life more expensive that kicked off the final death spiral of Trudeau's favourite tax amid collapsing support for the Liberals in their typical stronghold of Atlantic Canada.

CHAPTER 14

TRUDEAU'S ATLANTIC CANADA CONUNDRUM

The provincial patchwork of carbon taxes that Prime Minister Justin Trudeau allowed to flourish under the initial phase of his carbon tax scheme would later cause headaches. The short-term benefit of allowing some provinces to charge lower carbon tax rates in the early years created enormous challenges in the years to come.

Former Saskatchewan New Democratic Party finance minister Janice MacKinnon predicted regional tensions would tear apart a national carbon tax when she analyzed Liberal leader Stéphane Dion's proposed "Green Shift" carbon tax in 2008.

"What Dion was saying was that he was prepared to ignore the reality that the provinces also have jurisdiction over environmental issues and their cooperation would be essential in implementing his plan," MacKinnon wrote. "I tried to warn the Liberals of the political consequences of failing to address the regional unfairness of a national carbon tax."

MacKinnon's warnings largely focused on a national carbon tax "rekindling the Western alienation and regional discontent" that had been seen under the reign of Trudeau's father, Pierre Trudeau. And she was spot on in pointing out that any carbon tax imposed on the provinces from Ottawa would raise regional tensions throughout Canada.

While it was a safe bet that Trudeau's carbon tax would never be popular in the prairies, the prime minister failed to reckon with the possibility his carbon tax would also prove to be unpopular in other regions of the country, such as Atlantic Canada. And it was anti-carbon tax pressure from Atlantic Canadians that led to a major blow against the unpopular tax in the form of dissent among government MPs within the Liberal caucus.

To understand how the carbon tax became a political loser in Atlantic Canada, and ultimately across the country, we need to take a step back to the early days of Trudeau's carbon tax.

In theory, the federal carbon tax was a "national minimum" tax. That meant a province could impose its own carbon tax that met the federal requirements, or Ottawa would impose the federal carbon tax in the province. In practice, however, this resulted in a patchwork of different carbon tax rates across Canada.

The Canadian Taxpayers Federation (CTF) highlighted Trudeau's regional unfairness in our 2019 Gas Tax Honesty press conference, about a month after Trudeau's carbon tax took effect. At the time, the carbon tax was supposed to cost an extra 4.42 cents per litre of gasoline across the country, regardless of jurisdiction. In practice, the federal government allowed the carbon tax to cost one cent per litre of gas in Prince Edward Island, 0.42 cents per litre in Newfoundland and Labrador and less than one cent per litre in Nova Scotia.

Standing outside the Alberta Legislature, the CTF called out Trudeau for his carbon tax unfairness.

"It's bad enough that Albertans have been forced to pay more to fuel their cars and heat their homes during cold winters, but now Ottawa is adding insult to injury by letting other provinces pay a lower carbon tax," the CTF said.

At the press conference, the CTF displayed two large gas cans. One of the cans showed the carbon tax rate Albertans were being forced to pay: 6.7 cents per litre of gasoline in 2019, rising to eleven cents per litre in 2022. The other gas can showed the carbon tax rate Nova Scotians were paying: one cent per litre of gasoline.

And the CTF wasn't alone in noticing this disparity. Big-time political heavyweights also took notice of Trudeau's carbon tax unfairness following the press conference.

Danielle Smith, then a talk-radio host, now Alberta premier, praised the CTF's "good work exposing additional unfairness in how the federal government has levied carbon taxes on consumers." Smith added that "it looks as if the federal government is prepared to cut special deals for regions that vote Liberal and apply punitive regimes on regions that don't.

"This lack of coherence in policy is surely one of the reasons public support for consumer carbon dioxide taxes is vanishing."

The provincial patchwork of carbon taxes created a major problem for Trudeau and his Liberal government. To end the unfairness, the Trudeau government had three options: (1) scrap the carbon tax; (2) reduce carbon tax rates in the provinces that paid more; (3) increase the carbon tax on Atlantic Canadians.

The Trudeau government chose the worst option: increase the carbon tax in Atlantic Canada by imposing the federal carbon tax.

Nova Scotia's carbon tax sat at about two cents per litre of gasoline on June 30, 2023. The very next day, on July 1, Trudeau hiked the carbon tax on Nova Scotians to fourteen cents per litre. Overnight, Trudeau had effectively raised gas prices in the province by twelve cents per litre.

In the lead up to the Canada Day carbon tax hike in Atlantic Canada, the CTF did a media tour across the region, highlighting the coming federal carbon tax. Prior to Trudeau's tax hike, the CTF set up shop in downtown Halifax to host a press conference.

"This is a massive tax hike and Nova Scotians can't afford to pay more for gasoline and groceries because of a huge tax hike that won't help the environment," the CTF told Nova Scotian reporters.

At the press conference, the CTF had a massive poster showing what Trudeau's tax hike meant for Nova Scotians. The poster had two bar graphs: before federal tax hikes: two cents per litre of gas, after federal tax hikes: fourteen cents per litre of gas. At the bottom of the poster was a family standing beside their minivan with a price tag showing $431. That's because a Parliamentary Budget Officer report at the time showed the carbon tax would cost the average Nova Scotia family $431 more in 2023 than it got back in rebates.

Adding insult to injury for Nova Scotians was the fact the province had cut emissions by thirty-six percent since 2005—the second deepest cut in emissions among all Canadian provinces.

"Nova Scotia has been one of the leaders in Canada when it comes to reducing emissions and the province has done it without a punishing federal carbon tax," said Jay Goldberg, then the CTF's interim Atlantic director. "The Trudeau government should be looking to Nova Scotia for advice, not hammering the province with a massive tax hike that families can't afford."

The message the CTF sent across Atlantic Canada was clear: massive federal carbon tax hikes were coming, they would make life more expensive and the carbon tax doesn't work.

At the same time the CTF was doing its media tour across the region, Conservative Party leader Pierre Poilievre was hosting massive Axe the Tax rallies across the region.

"Who's ready to axe the tax?" Poilievre asked a cheering crowd full of hundreds of supporters in Truro, Nova Scotia. Poilievre was able to get hundreds of people out to a rally more than two years away from the next scheduled federal election, in a town of less than 13,000 people, based on the message of axing the carbon tax.

The Atlantic Chamber of Commerce also rallied against the carbon tax. "The new carbon taxes deliver a one-two punch in Atlantic Canada," the chamber said in a press release.

Clearly, the message of the carbon tax making life more expensive was breaking through in Atlantic Canada. Typically, taxes aren't a major topic of conversation. But leading up to the 2023 carbon tax hike in Atlantic Canada, people were talking about the carbon tax hike in grocery stores, restaurants and even at the airport.

"I heard about the new carbon tax and it's absolutely ridiculous," Nova Scotia delivery driver Emran Hassan told *CTV*. "It's like we're playing a game of survival."

"Ludicrous" is how Islander Janet Vanderkay described the carbon tax hike to *CBC*, especially when it came to home heating oil. "Honestly, I don't know how they're expecting us to survive."

The pushback from Atlantic Canadians to Trudeau's carbon tax hike was instantaneous, resulting in a significant blow to the Liberal government.

The Atlantic Canadian premiers launched their Fight the Federal Gas Hike campaign in response to the federal carbon tax hike, which came at the same time as the imposition of the second carbon tax through fuel regulations (also July 1, 2023).

"Nova Scotians are concerned about the clean fuel regulations," Nova Scotia Premier Tim Houston said during the campaign launch. "Combined with the carbon tax this will increase the cost of everything—fuel, food, clothing and more."

"We are committed to finding a balanced approach that addresses climate change while ensuring affordability—but innovation and improved technology are what will get us there, not more taxes," New Brunswick Premier Blaine Higgs said. "The impact of these additional costs extends beyond the pumps and heating our homes—they affect our grocery bills and the daily necessities we rely on."

Prince Edward Island Premier Dennis King, who served as the Chair of Atlantic Premiers Council, outlined how carbon taxes were especially harmful to the region due to its geographic realities.

"The federal Clean Fuel Regulations unfairly affect Atlantic Canadians because of a variety of factors including our limited fuel sources, a lack of major transit systems, our system of trucking in required goods, and residents with less financial flexibility to bear additional costs or make different choices," King said.

But the sharpest dagger came from a Liberal: Newfoundland and Labrador Premier Andrew Furey.

Calling into Newfoundland and Labrador's political talk radio show, Voice of the Common Man, Furey ripped Trudeau for his claim that you either supported the federal carbon tax or were a climate change denier. Furey called this "completely illogical" and "as insulting to us as it is simplistic."

The backlash was swift. Polling numbers for the Trudeau Liberals in the region, which had long been a stronghold for the party, dropped. Before the federal carbon tax hike, the Liberals had been polling at 38.3 percent in Atlantic Canada. By October 22, they were polling at 33.3 percent and were overtaken by the Conservatives, according to CBC's poll tracker.

"There has been a significant backlash against the levy in Atlantic Canada," reported the *Globe and Mail* in October 2023. "Federal carbon pricing took effect on the East Coast in July. Since then, local MPs have received an earful about the higher costs. Opinion polling shows the Conservatives have gained a double-digit lead over the Liberals in the region."

Not only were the federal MPs getting "an earful" over the higher costs the carbon tax imposed on Atlantic Canadians, frustration within the region spilled into provincial politics.

Following the federal Liberals' carbon tax hike, the provincial Liberals lost a seat in a by-election that the party had held for twenty years. In fact, the provincial Liberals finished third in that by-election, behind the Progressive Conservatives and the provincial

New Democrats. The PCs heavily criticized the "Liberal carbon tax" during the campaign.

By the fall of 2023, the seeds of a carbon tax revolt had been planted and watered within the federal Liberal caucus. Ken McDonald, a Liberal MP from Newfoundland and Labrador, voted on October 4, 2023, in support of a Conservative motion to "repeal all carbon taxes."

"Everywhere I go people come up to me and say, 'You know, we're losing faith in the Liberal Party,'" McDonald told *CBC*. "Seniors who live alone tell me they go around their house in the spring and winter time with a blanket wrapped around them because they can't afford the home heating fuel."

Soon after, three Liberal MPs from New Brunswick—Serge Cormier, René Arseneault and Wayne Long—issued a joint press release raising concerns over the carbon tax. The carbon tax "puts the province and its people at a disadvantage because it does not take into account the fact that it is largely rural," the Liberal MPs said.

Then, on October 26, 2023, a political bombshell went off. That afternoon, my phone vibrated and I read a vague email from *CTV National*: "The prime minister is expected to make an announcement at 4 p.m. that pertains to affordability for Canadians. Would you be able for a brief recorded Zoom interview around 4:30 or 5 p.m.?"

It was around 3:30 p.m. A major announcement from Trudeau on affordability. I couldn't help but wonder: was Trudeau about to announce the end of the carbon tax? A few minutes later, my phone vibrated again. It was another email from CTV: "We believe there may be environmental impacts."

Trudeau took the mic, surrounded by Liberal MPs from Atlantic Canada, and announced the federal government would remove the carbon tax from home heating oil for three years—a fuel source primarily used in Nova Scotia and Prince Edward Island.

Unfortunately, on that day, Trudeau didn't scrap the carbon tax. In fact, the vast majority of Canadians would see no tax relief from the move. But there was something critical about the announcement that I didn't immediately grasp. And a hint was staring me right in the face, in the slogan on Trudeau's podium: "Making Life More Affordable."

The Trudeau government had consistently claimed most Canadians got more money back in rebates than they paid in carbon taxes. In fact, Environment Minister Steven Guilbeault rarely missed an opportunity to proclaim the "Carbon Rebate gives more back to eight out of ten Canadians."

But if the carbon tax wasn't making life more expensive, then why had Trudeau just provided relief to Atlantic Canadians by removing it from home heating oil? And why would his podium carry a slogan about making life more affordable? With this carve-out, Trudeau had all but publicly admitted the carbon tax did, in fact, make life more expensive.

The carve-out also proved the carbon tax had a lot more to do with politics than it did with saving the planet. It took one Liberal MP from Atlantic Canada to show conviction and push back against the carbon tax for Trudeau to cave. Atlantic Canada, after all, is typically a Liberal stronghold. Given the fact that Trudeau's poll numbers were slumping, it was not a region he could afford to lose.

By announcing his carbon tax carve-out surrounded by his "amazing" Atlantic Liberal MPs, Trudeau had given the game away. Others extended his logic. "Perhaps they need to elect more Liberals," said Gudie Hutchings, minister of the Atlantic Canada Opportunities Agency, when asked why most Canadians weren't getting any carbon tax relief.

The unfairness was infuriating for millions of Canadians. While up to forty percent of Atlantic families used home heating oil, only three percent of Canadians did. In other words, Trudeau's carbon tax carve-out left ninety-seven percent of Canadian families out in the cold.

On that day, Trudeau didn't end the carbon tax. But with the benefit of hindsight, it's clear that on October 26, 2023, the day Trudeau stood at that podium and announced his carve-out, the end of the carbon tax was in sight.

CHAPTER 15

THE UNRAVELLING OF THE CARBON TAX

"The carbon tax may soon be dead," wrote economist Trevor Tombe, a vocal supporter of carbon taxes, in October 2023. "And we'll have the heating oil exemption to thank."

Prime Minister Justin Trudeau's carve-out for Atlantic Canada dealt a major blow to the federal carbon tax. It undercut the government's claim that most households are left better off with its carbon tax and rebates. It also showed the carbon tax was all about politics, not helping the environment. Soon enough, the whole carbon tax scheme would begin to unravel.

Following the carbon tax carve-out, the *Globe and Mail* wrote an editorial entitled "The Liberals' credibility on the carbon tax has gone up in smoke."

"In reality, the answer is this: those regions with the prescience to deliver Liberal seats will get help," the *Globe and Mail* wrote, describing the carve-out as "rank regional favouritism."

And this was the biggest blow to the carbon tax: the carve-out united the country against it.

That's because the Liberal government decided to give carbon tax relief to three percent of families, while leaving ninety-seven percent of Canadians out in the cold paying the carbon tax in full for the next three winters.

Or, as Alberta Premier Danielle Smith put it, "The federal government has decided that one part of Canada with one type of home heating is worthy of a carbon tax break, while those living elsewhere using another type of home heating do not."

But it wasn't just one Conservative premier from Alberta speaking out. Premiers across the country and across the political spectrum spoke out against Trudeau's ploy.

Shortly after the carbon tax carve-out was announced, five premiers representing more than half the population of Canada called for "a meeting with the prime minister on carbon tax fairness for all Canadian families."

"It is of vital importance that federal policies and programs are made available to all Canadians in a fair and equitable way," the premiers said in a joint statement. "By singling out Atlantic Canadians with this relief, it has caused divisions across the country. All Canadians are equally valued and should be equally respected."

Even New Democrat premiers were speaking out.

"At a minimum, fairness demands equal treatment of British Columbians," BC's New Democratic Party (NDP) Premier David Eby said. "People struggling with affordability around home heating face the same struggle in BC It's not a distinct or different struggle."

"The carbon tax is not a silver bullet when it comes to climate change," Manitoba's NDP Premier Wab Kinew said, adding that the feds should "help people who are struggling right now because of inflation."

In Ottawa, the federal NDP voted in favour of a motion to extend the carve-out and remove the carbon tax from everyone's home heating bills.

"The reality is we need to make sure that affordability is available to all Canadians and that's why we're supporting this motion," said federal NDP house leader Peter Julian.

Even the Alberta NDP, which had brought in a surprise provincial carbon tax, criticized Trudeau's carve-out.

Alberta NDP leader and former premier Rachel Notley said applying the carbon tax to "some regions, and some fuels, but not all, is totally unacceptable."

Sarah Hoffman, who had been Notley's deputy premier, took it a step further when she ran to be the new Alberta NDP leader.

"I think the consumer carbon tax is dead," Hoffman said. "It died provincially in the last election. The feds took it over. Justin Trudeau played dirty politics with it and picked winners and losers. If you don't have public support, you can't carry on with something like that."

Here's the key takeaway: following Trudeau's carve-out, politicians of every political stripe saw the writing on the wall and spoke out in opposition to the federal carbon tax. Even more damaging, Trudeau's home heating oil carve-out didn't quell opposition to the carbon tax in Atlantic Canada—the very place it was meant to provide relief.

Less than one month after Trudeau's carve-out, the Canadian Taxpayers Federation (CTF) released Leger polling revealing what Atlantic Canadians thought of the home heating oil exemption.

The poll showed seventy-seven percent of Atlantic Canadians wanted the carbon tax to be removed from everyone's home heating bills. Trudeau tried to buy off MPs in the region with his carbon tax carve-out, but Atlantic Canadians were demanding relief that was fair for everyone.

Even Canada's lone Liberal premier continued to swing away against the carbon tax. After Trudeau removed the carbon tax from home heating oil, Newfoundland and Labrador's Andrew Furey continued to call for the carbon tax to be removed from other fuels.

"There is no subway in St. Anthony," Furey said. "You do have to drive a (pickup truck) if you're a crab fisherman hauling crab pots, or a miner or a logger … so there are no options for change."

Furey went on to say that because so many people have no other option but to fuel up their cars with diesel or gasoline, "the fundamental premise on which the whole [carbon tax] thing is based, is flawed."

Following the home heating oil carve-out, premiers in the prairies decided to take matters into their own hands.

Saskatchewan Premier Scott Moe chose to fight the carbon tax in a new way, by removing it from home heating in the province.

The Moe government directed SaskEnergy, the province's natural gas Crown corporation, to stop charging the carbon tax on natural gas bills, saving Saskatchewan families about $400 in 2024 and an estimated $480 in 2025.

"Our government is ensuring fairness for Saskatchewan families by removing the federal carbon tax on natural gas and electric heat, just as the federal government has done for families in Atlantic Canada by removing the carbon tax on heating oil," Saskatchewan Crown Investments Corporation Minister Dustin Duncan said. "By extending carbon tax relief to Saskatchewan families who were left out in the cold by the federal government, our government is protecting Saskatchewan families' ability to afford to heat their homes this winter."

Importantly, Moe's gambit to stop collecting the carbon tax on natural gas bills was supported by the Saskatchewan NDP.

In Alberta, Smith launched a renewed court challenge against the federal carbon tax. Here's what Smith said when she announced Alberta was taking the federal carbon tax back to court:

> Ottawa decided Canadians in the East deserved a three-year break from paying the carbon tax on their home heating costs. While we're happy for these Canadians, Alberta, Saskatchewan and other provinces who heat their homes with natural gas have been deliberately excluded from these savings. Albertans simply cannot stand by for another winter while the federal government picks and chooses who their carbon tax applies to. Since they won't play fair, we're going to take the federal government back to court.

It's easy to understand the frustration felt in Western Canada. While furnace oil makes up to forty percent of home heating in the Atlantic Canadian provinces, it only makes up one percent in British Columbia and less than one percent in Alberta, Saskatchewan and Manitoba.

Other than the federal Liberals, politicians of all political stripes were fighting back against the carbon tax. To put it mildly, support for the carbon tax had collapsed. This is captured perfectly in a 2024 poll released by the Angus Reid Institute, published shortly before the government again hiked its carbon tax on April 1. Nearly eighty percent of Canadians wanted the federal government to abolish, cut or freeze the carbon tax, according to the poll.

Let that sink in: almost eighty percent of Canadians were either against the carbon tax outright, or were against the government's plan to keep cranking up the tax. If that wasn't bad enough, the poll also showed Canadians weren't buying the government's two biggest talking points on the carbon tax.

Canadians didn't think the carbon tax worked.

"Criticism of the carbon tax is driven by a sense that it is ineffective at reducing Canada's greenhouse gas emissions (sixty-eight percent say this)," according to the Angus Reid Institute.

The problem for the Trudeau Liberals was that Canadians knew we accounted for a minuscule amount of global emissions—just 1.4 percent, according to government data. And Canadians can look south of the border to the world's largest economy and our biggest trading partner and competitor, and see that the US government doesn't impose a carbon tax. In fact, about seventy percent of countries don't impose national carbon taxes, including four of the five largest emitters.

Canadians can also look at government data to see that carbon taxes don't work. BC imposed Canada's first broad-based carbon tax in 2008. Politicians promised the carbon tax would cut 2007 emissions by thirty-three percent by 2020. But the government's own data shows

BC's emissions went up. The federal government own records show it doesn't know how much of Canada's emissions are cut by the carbon tax, and the environment commissioner's latest report shows the government is far from meeting its promised targets.

Second, Canadians know the carbon tax makes life more expensive.

Just four percent of Canadians say the carbon tax and rebate scheme saves them money, according to the Angus Reid poll. Meanwhile, forty percent of Canadians say the carbon tax is making their life "a lot" more expensive, while twenty-six percent say it's increasing their cost of living "a little." And that's not the only poll showing broad opposition to the carbon tax.

Just before the 2024 carbon tax hike, the CTF released a Leger poll showing seven-in-ten Canadians opposed the increase. The majority of Canadians across all demographics—gender, age, province, income, education level—opposed the carbon tax hike.

Across the political spectrum, politicians swung away against the carbon tax hike.

The Conservatives rode a fifteen- to thirty-point lead in the polls through all of 2024 as they campaigned to Axe the Tax.

The only Liberal premier, Newfoundland and Labrador's Andrew Furey, published an open letter calling on the Trudeau government to drop its carbon tax hike.

"I continue to stand up for Newfoundlanders and Labradorians against the federal carbon tax," Furey said. "I am now asking Ottawa to pause its planned increase to the carbon tax, set for April 1, as the high cost of living is enough of a burden on families."

Through a vote in the provincial legislature, Nova Scotia's Progressive Conservatives, Liberals and New Democrats all called on Prime Minister Justin Trudeau to cancel the increase.

"Nova Scotians will pay over twenty-three cents in carbon taxes per litre at the gas pump, with a Liberal carbon tax increase of 3.3 cents per litre," the text of the vote read. "All members of the Nova Scotia

Legislature call on Nova Scotia MPs to vote in the best interests of Nova Scotians by standing up against an increase in the carbon tax."

Despite widespread opposition to the carbon tax from Canadians and politicians across all parties, the Trudeau Liberals went ahead with their carbon tax hike, triggering protests across the country.

"Everything you purchase, every one of these trucks going by, every one of these cars going by, everything we consume in this country is going up today," a protestor in Alberta said. "Thanks Justin, you just made things even less affordable for Canadians."

To illustrate just how unpopular the carbon tax has become, consider the case of British Columbia.

The BC Liberals introduced the first broad-based carbon tax in Canada in 2008, which was a move praised by many academics, environmental activists, international organizations, media elites and politicians. The *New York Times* said BC's carbon tax could be "a template for the rest of the world." *The Economist* referred to BC's carbon tax as "a winner."

But by 2024, it had become clear British Columbians thought the carbon tax was a loser. Leger polling commissioned by the CTF showed six-in-ten British Columbians wanted the BC government to immediately scrap its provincial carbon tax. Only twenty-five percent of British Columbians supported the carbon tax, according to the poll. And when you removed people who were undecided on the issue, seventy percent of British Columbians said the province should immediately scrap the provincial carbon tax.

Prior to the last provincial election in 2024, the BC United Party—formerly the Liberal Party, which had introduced BC's carbon tax—reversed course and said it would scrap the carbon tax. The newly resurgent BC Conservative Party gained ground in the polls by attacking the carbon tax. Even NDP Premier David Eby said he would cancel the provincial carbon tax if the federal government scrapped its mandatory minimum.

"A lot of British Columbians are struggling with affordability," Eby said. "Our commitment is that if the federal government decides to remove the legal backstop requiring us to have a consumer carbon tax in British Columbia, we will end the consumer carbon tax in British Columbia."

Outside of Liberals in Ottawa, politicians of all political stripes are now either opposing the carbon or have spoken out against it. And that's because the people oppose the carbon tax.

As former Alberta premier Ralph Klein would say, the key to political success is to find out what the people want and get in front of the parade. The politicians weren't leading. They were following the people's opposition to the carbon tax in the hope of keeping their jobs.

And while Trudeau's carbon tax carve-out may appear to have been the beginning of the end, the truth is that Canadians are fed up with carbon taxes in general.

Yes, Trudeau stepped in it when he removed the carbon tax from home heating oil, while leaving ninety-seven percent of Canadians out in the cold. Even in Atlantic Canada, where Trudeau tried to buy off MPs with the carve-out, seventy-seven percent of people in the region support carbon tax relief for everyone.

But Trudeau's mistake wasn't providing relief in Atlantic Canada. The real lesson here is that Trudeau never won the hearts and minds of Canadians.

He wasn't elected prime minister because he ran on a carbon tax. In fact, he barely mentioned his carbon tax plans during the 2015 election. And he lost credibility early on.

Months before the 2019 election, former environment minister Catherine McKenna said the government had "no intention" of raising the carbon tax beyond eleven cents per litre of gas. After the election, Trudeau announced he would keep cranking up his carbon tax until it reached thirty-seven cents per litre.

Trudeau and his ministers repeated the myth that eight out of ten families get more money in rebates than they pay in carbon taxes. Their favourite talking point limps on despite the obvious reality that a government can't raise taxes, skim money off the top to pay for hundreds of administration bureaucrats, charge its sales tax on top and somehow make everyone better off. In fact, the carbon tax costs average families hundreds of dollars more every year than they get back in rebates, according to the PBO.

The government said carbon taxes reduce emissions. But even in BC, which had the first and (for years) costliest carbon tax in Canada, emissions rose. And for the sake of argument, even if the carbon tax did cut emissions at home, "Canada's own emissions are not large enough to materially impact climate change," as the PBO explains.

Making it more expensive to live in Canada won't reduce emissions in China, Russia, India or the United States. And this leads us to Trudeau's diplomatic failure. At the United Nations, the Trudeau government launched the Global Carbon Pricing Challenge to get more countries to impose carbon taxes.

"The impact and effectiveness of carbon pricing increases as more countries adopt pricing solutions," the Trudeau government has acknowledged.

The world's largest economy, the United States, rejects carbon taxes. Democrats occupying the White House haven't imposed a carbon tax. Good luck convincing a Republican president to impose one. And the US is the rule, not the exception. About seventy percent of countries *don't* have a national carbon tax, according to the World Bank's Carbon Pricing Dashboard.

And while Trudeau raised taxes, peers like the United Kingdom, Sweden, Australia, South Korea, the Netherlands, Germany, Norway, Ireland, India, Israel, Italy, New Zealand and Portugal, among others, cut fuel taxes. Not to mention provincial governments of all political stripes, like New Democrats in BC and Manitoba, Conservatives in

Alberta and Ontario, and Liberals in Newfoundland and Labrador, provided fuel tax relief.

Then there's the carbon tax failure when it comes to fairness. If Canada's carbon tax is essential for the environment, shouldn't all taxpayers pay the same rate?

A driver in Alberta pays a carbon tax of seventeen cents per litre of gas. In Quebec, the carbon tax is about twelve cents. By 2030, that gap will grow to more than fourteen cents per litre. As that gap grows, Ottawa will face explosive pressure to step in and crank up Quebec's carbon tax, as it did in Nova Scotia in 2023. If Ottawa couldn't handle political backlash from Atlantic Canada, what's was the plan when it's coming from Quebec?

Quebec's special deal proves Trudeau's carbon tax is about politics, not the environment.

When crafting the carbon tax, the government never truly asked the people what they thought. Everyone wants a better environment. You won't find opposition to that.

But did anyone ask Canadians if they supported a carbon tax even if it meant average families would lose hundreds of dollars every year? Did anyone ask Canadians if they supported a carbon tax even though most countries won't impose one?

Trudeau displayed regional favouritism with his carbon tax carve-out. But his real mistake wasn't a carve-out that favoured Atlantic Canada. The real failure is that he never convinced Canadians of his plan and failed to acknowledge the carbon tax would cause real pain.

CHAPTER 16

THE CARBON TAX FIGHT AHEAD

Le Roi est mort, Vive Le Roi!

"The king is dead, long live the king!" was first declared in 1422 when King Charles VI died and his son King Charles VII ascended to the French throne.

It looks like the federal carbon tax is on its deathbed. But those of us who oppose carbon taxes can't let up, because proponents of the carbon tax are already working to revive it.

It's hard to not be giddy when looking at the polls. A Leger poll conducted for the Canadian Taxpayers Federation (CTF) found seven-in-ten Canadians opposed the government's most recent carbon tax hike. Another poll from the Angus Reid Institute found nearly eighty percent of Canadians want the carbon tax abolished, cut or at the very least not increased.

Pierre Poilievre's Conservatives are running an Axe the Tax campaign and they've been anywhere between fifteen to thirty points up in the polls. In effect, the person likely to win the next election has made scrapping the carbon tax his central promise to voters.

Eight-of-ten provincial premiers pushed back against the latest federal carbon tax hike. This included the only remaining Liberal premier,

Andrew Furey of Newfoundland and Labrador, and Manitoba NDP Premier Wab Kinew. And with Prime Minister Justin Trudeau resigning, Liberal MPs say the carbon tax "clearly needs to be re-examined."

Given all this momentum, it's tempting to dance on the carbon tax's grave. But even if the carbon tax is finally killed in 2025, carbon tax proponents will be trying to revive and disguise the monster.

After former Liberal leader Stéphane Dion's crushing defeat at the polls in 2008, his Green Shift carbon tax looked dead. But it wasn't. It was hibernating.

And the carbon taxers are already circling the wagons.

The Ecofiscal Commission launched an open letter signed by economists begging the government to keep the carbon tax. Left-wing pundits are calling for the Liberal government to kill the carbon tax now, so the 2025 election isn't fought and lost on a carbon tax, thereby poisoning the well for the future. Undoubtedly, in the bowels of the David Suzuki Foundation, someone is working on renaming, re-framing and re-working a carbon tax into an "industrial price on pollution" or a complicated cap-and-trade system. But the end result will still make heating your home and driving your car more expensive.

For Canadians, there are two important things to remember.

First, it's important to stay vigilant. Even if the government kills the carbon tax, the fight will not be finished. The carbon taxers will regroup to bring back a form of carbon tax that bites with a vengeance.

Second, a carbon tax is a carbon tax is a carbon tax. No matter how many times academics, the media or environmental ideologues try renaming it—whether it's an industrial carbon tax, cap and trade or some type of environmental levy—the simple fact remains that carbon taxes make life more expensive and don't work to help the environment.

The Need to Stay Vigilant

No defeat is forever and no victory is everlasting.

That's especially true in politics. If Canadians want to rid themselves of carbon taxes, it will take vigilance to beat back every carbon tax gopher that pops up. To understand this lesson in the fight ahead, look no further than Canada's recent history with carbon taxes.

Despite carbon taxes being defeated at the polls and in the eyes of public opinion, most significantly in 2008 with Dion's Green Shift, carbon tax proponents have not given up pushing the policy throughout the years. A quick search of Google Scholar for "Canada Carbon Tax" shows more than 300,000 results.

The David Suzuki Foundation claims it has been pushing for a carbon tax since 1998. In that same year, Prime Minister Jean Chretien signed Canada onto the Kyoto Protocol—an international climate change agreement that called for cap-and-trade carbon taxes.

After the Liberal's crushing defeat in the 2008 election, some columnists still called for some form of a carbon tax in the future. Andrew Coyne claimed that it wasn't the carbon tax itself that killed Dion's chances of winning, but rather how he communicated the policy.

Sound familiar? Trudeau's carbon tax wasn't the problem, it was poor communication.

"The best product in the world still needs a good salesman," Coyne wrote in 2008. "A complicated new product needs a really good salesman. Dion did not sell this at all well."

Canada's Ecofiscal Commission, a group of academics, started publicly calling for a carbon tax in 2015. Trudeau then revived the carbon tax that had been devastatingly defeated in 2008.

Or take the Australian example, where the carbon tax was defeated and then quickly brought back to life. Under former Labor Prime Minister Julia Gillard, the Australian government imposed a carbon tax on the country's largest emitters in 2012.

"It cost the Australian economy $8 billion a year for two years, it raised electricity prices by twenty-five percent … and contributed to higher prices at the supermarket," said Chris Berg, senior fellow with Australia's Institute of Public Affairs.

Australians were told by the government that a carbon tax was needed to "meet climate change obligations of Australia." Like Canada, Australia accounted for about 1.5 percent of global emissions at the time.

"The idea that anything we in Australia could do to make a legitimate impact on climate change was fairly ludicrous," Berg said. "It's easy to say we're making big polluters pay, but fundamentally these taxes have an effect on individual people—people have to pay these taxes."

After the introduction of the tax, opposition leader Tony Abbot made a "pledge of blood" and promised to "axe the tax." Abbot made good on his promise and repealed the carbon tax in 2014. A victory for Australian taxpayers. But a short-lived one.

After axing the tax, the government designed a new carbon tax scheme called the safeguard mechanism, which was implemented in 2016. This scheme applied to companies and if they emitted over a certain amount, they would either have to surrender "carbon credits" to the government, earned by doing projects that reduce or sequester emissions, or pay fines.

At the outset, the government did not strictly enforce it. But fast forward a couple of years and the Australian government reformed the safeguard mechanism in 2023. The reforms mandate that all the companies covered must start to reduce their emissions every year. If the companies fail to reduce their emissions, they can use their carbon credits or pay for every tonne of emissions over the amount set by the government.

If that sounds like a carbon tax, that's because it's a carbon tax.

Both the Canadian and Australian example have a clear takeaway for taxpayers: we must stay vigilant. A carbon tax could be scrapped today, but the carbon tax activists will regroup and try to force it down Canadians' throats again tomorrow.

And just like in Australia, the thing to keep an eye out for is a carbon tax by another name.

Fighting a Carbon Tax by Another Name

A carbon tax is a carbon tax is a carbon tax.

When Trudeau introduced his "national minimum" carbon tax it included two parts. The first was a consumer carbon tax directly at the gas pumps and on your heating bill. But there's also Trudeau's hidden, industrial carbon tax on businesses, like fertilizer plants and oil and gas companies.

The hidden carbon tax works in a similar way as the consumer carbon tax. Either the provinces impose their own industrial carbon tax that meets the federal requirement, or the federal industrial carbon tax is directly imposed on those provinces.

That hidden carbon tax also impacts everyday Canadians. That's because when the government hits businesses with a carbon tax, they don't just eat the extra costs, they pass it onto consumers through higher prices or onto workers by cutting back production or shifting it to other jurisdictions, like the United States, which don't impose industrial carbon taxes.

Writing about the industrial carbon tax in 2019, the PBO noted "the cost of the GHG emission is 'passed through' to the final consumer of the product through an increase in the sale price."

To his credit, Poilievre has been a vocal opponent of the consumer carbon tax, promising to "axe the tax" and save Canadians money at the pumps and on their heating and grocery bills. But Poilievre has been less vocal about Trudeau's industrial carbon tax, an issue that multiple media outlets have reported on.

"Pierre Poilievre says he'll 'axe the carbon tax' for consumers—but when it comes to the pricing system for Canadian industries, he won't say," reads a headline from the *Toronto Star*.

Worse still, environmental activists, academics, industry groups and even former Conservative Party strategists have started putting pressure on Poilievre and his MPs to keep some form of carbon tax, including the industrial carbon tax.

"The debate around the consumer carbon [tax] is sucking too much oxygen out of the room," said Rick Smith, president of the Canadian Climate Institute. "Industrial carbon pricing is the most important contributor to carbon reduction across the country."

University of Victoria professor Peter Dietsch wrote an opinion piece praising the notion of carbon taxes. The title of his article was, "The carbon tax needs fixing, not axing."

Or take the *Globe and Mail* editorial entitled "Why Canada needs better carbon taxes," which claimed "the biggest opportunity to boost carbon pricing is in its industrial incarnation." The piece concluded by calling on federal politicians to "refocus their climate policy on improving, not subsuming, carbon taxation."

Max Fawcett, a blogger at the *National Observer*, an outlet that publishes "on the defining crisis of our time: climate change," had a more direct call to action for carbon tax activists. Fawcett called for a "full strategic retreat" and wants to see "the consumer portion of the carbon tax" scrapped while the "the industrial carbon tax" is maintained. In effect, Fawcett wanted the Trudeau government to "kill the carbon tax and live to fight another day."

Big business has also pushed for Ottawa to keep its hidden carbon tax.

"The Chemical Industry Association of Canada (CIAC), say they want industrial pricing to stay in place, but also want to see additional support like the promised tax credits for green investments," the *Toronto Star* reported.

"Our message is to continue with carbon pricing," CIAC president Bob Masterson told the *Toronto Star*.

And here's the cherry on top: "But he said more support will be needed if the price keeps going up as the Liberals have planned," the *Toronto Star* reported.

Do you see how this works? Some big industry lobby groups don't mind a hidden carbon tax that will limit smaller competitors from entering the market and whose cost will ultimately be, at the very least in part, passed onto Canadian consumers, as long as they get taxpayer subsidies from the government.

Former Conservative Party strategist Ken Boessenkool has repeatedly implied that even a government under Poilievre, who is running on an Axe the Tax campaign, may not scrap the carbon tax completely.

"If you listen closely, the Liberal's biggest threat—that would be Pierre Poilievre—has been campaigning hard on eliminating the carbon tax," Boessenkool wrote in 2023. "He's not always suitably precise, but I don't actually think he plans on eliminating the entire carbon tax. What he intends to do is eliminate the retail side of the carbon tax. Conservatives have supported industrial carbon taxes when given the chance."

Boessenkool points to former Alberta premier Jason Kenney, who "campaigned against the carbon tax," but whose "government still collected the industrial carbon tax during his time."

"Now it's entirely possible that both the industrial and retail carbon taxes will disappear if/when Poilievre moves into whatever dilapidated residence the prime minister gets to live in these days," Boessenkool said. "But I don't think so. I think we are on a path that lifts, reduces or eliminates the carbon tax on the retail side, but maintains it on the industrial side."

Boessenkool concluded his piece with: "So the (retail) carbon tax is dead. Long live the (industrial) carbon tax."

This issue pops up with other politicians who promised to scrap the carbon tax.

British Columbia Premier David Eby said he would scrap the consumer carbon tax if the federal government removed its national minimum backstop.

"We'll remove the carbon tax for everyday British Columbians, for the farmers, for the truckers, for the average British Columbian," Eby said. "We will continue to ensure … that the big polluters are paying their fair share."

Another scheme Canadian opponents of carbon taxes must watch out for is a change in name, like what happened in Australia, or like what occurs in Quebec, where the provincial government imposes a cap-and-trade carbon tax.

With cap and trade, the government sets up an artificial market and mandates a certain level of emissions that decline over time. The government-mandated emissions cap determines the carbon tax a company must pay to buy credits. The key thing to understand is that a cap-and-trade scheme is just a carbon tax by another name. It may be called something different, but it still results in higher gas, diesel and home heating costs, along with reduced jobs and incentives for businesses to set up shop in other countries.

There's one other danger Canadians need to consider. While cap and trade, or a hidden, industrial carbon tax, or any other form of carbon tax, will still make life more expensive and make Canada less competitive, they are less transparent and harder for voters to sniff out.

"Politically, cap-and-trade has functioned as a 'safe harbour' for politicians who grasp the need to price carbon emissions but cling to the need to 'hide the price' to appease interest groups and/or voters," the Carbon Tax Center notes.

Back in Canada, a *Globe and Mail* editorial let the cat out of the bag by claiming it's better to have an industrial carbon tax that "is not as visible to irate voters." The media elites seem to think we should have carbon taxes, just one those pesky "irate voters" can't see.

These are three key takeaways for Canadians opposed to carbon taxes:

First, Canada has a number of carbon taxes, including the hidden, industrial carbon tax. And while the consumer carbon tax seems to be on its last legs, that's not necessarily the case for the others.

Second, regardless of its name, a carbon tax will make life more expensive for Canadians and won't move the needle on global emissions. This is still the case for a carbon tax on big business. A carbon tax on fertilizer producers is a tax on all Canadians that buy food. A carbon tax on gasoline refiners is a carbon tax on all Canadians that buy gasoline. A carbon tax on business will increase prices and lower the number of Canadians employed by businesses.

Third, carbon tax activists are already pushing for a "full strategic retreat" away from the consumer carbon tax, only to redouble their efforts pushing the hidden, industrial carbon tax or a carbon tax by another name.

Push for Carbon Taxes Come from Every Angle

It's not just environmental activists, academics and corporate lobbyists in Canada that we need to watch. International elites are looking for ways to impose global carbon taxes. In fact, international organizations are drooling over carbon taxes and Canada's current environment minister thinks the idea sounds swell.

"We are very supportive of the discussions that are happening at the International Marine Organization to put in place some kind of levy on international marine transportation," Environment Minister Steven Guilbeault said at the United Nations' recent annual climate conference in Baku, Azerbaijan, in November 2024.

Guilbeault doesn't pay the carbon tax when he burns through jet fuel and tax dollars flying around the world to climate conferences.

But he seems to support a tax on the shipping sector that bring things like phones and bikes to Canadians.

The United Nations' International Marine Organization wants "a maritime GHG emissions pricing mechanism." Translation: a new carbon tax on everything shipped across the oceans, like the shoes, cars, clothes and appliances Canadians buy, or the wheat, energy and minerals Canadians sell abroad.

The International Monetary Fund (IMF) recently published a report on "the urgent need for a global carbon tax on aviation and shipping."

The irony of politicians flying around the world on the taxpayer dime imposing a carbon tax on citizens when we fly appears lost on the IMF.

The IMF is at least up front about the costs, noting that a global carbon tax on aviation and shipping fuels would be "mostly passed through into flight ticket prices and shipped products."

This global carbon tax would cost the world's taxpayers "$200 billion by 2035, which could make a substantial contribution to climate finance for developing economies," according to IMF estimates.

In other words, your kids' shoes and family vacations will be more expensive. Then the unelected and unaccountable bureaucrats at international organizations will dictate which country will take your carbon-taxed cash.

In Canada, a global carbon tax on international flights and shipping would cost about 0.52 percent of GDP by 2035, or a $23.5 billion hit to our economy, according to the PBO.

Let's translate that to make sense for real people: this global carbon tax would cost a Canadian family of four about $2,000.

The UN and IMF aren't the only ones pushing for carbon taxes.

The World Economic Forum (WEF) published an article in January 2025 on "how to build a multilateral carbon pricing system," where it laid out a roadmap "toward a global carbon pricing mechanism."

This follows the 2024 annual WEF conference in Davos, Switzerland, where "world leaders heard a call for a global carbon tax as a solution to climate change during a panel discussion."

"There is no realistic solution to the climate transition that does not involve a globally coordinated system of carbon taxes," Singapore President Tharman Shanmugaratnam said at the WEF panel.

At least one politician knew his taxpayers back home wouldn't go for a global carbon tax.

Christian Lindner, Germany's finance minister, responded, "I hope that nobody tells my coalition partners and my colleagues and cabinet that we are here considering raising taxes. It would cause huge problems for me domestically."

The World Trade Organization has also pushed for global carbon taxes, as have international media elites. The *Financial Times* wrote an editorial entitled "The path to global carbon pricing."

There's one more push for carbon taxes coming from an international angle that Canadians should know about.

"The debate over the carbon tax has focused so far on domestic politics," wrote Eric Van Rythoven, an instructor at Carleton University. "However, this misses the importance of the international context. Increasingly, our trading partners take the threat of climate change seriously and use carbon tariffs to punish other countries they see as free-riders."

Van Rythoven continued: "Any government that wants to protect Canada from these tariffs will need a credible plan to reduce emissions. The result is that a future Conservative government may have to bring back the carbon tax, whether it likes it or not."

Essentially, the argument is that the federal government must impose carbon taxes on Canadians or some Canadian businesses will be forced to pay carbon taxes when they sell their goods to other countries, including those in the European Union.

This argument falls flat following a minute of critical thinking.

Imposing harmful carbon taxes on all Canadians due to the chance *some* Canadian businesses will pay carbon tax tariffs when they sell *some* of their products to *some* countries doesn't make sense. Why would the government inflict a tax on all Canadians—including all Canadian businesses—because there is a chance that some businesses will pay a carbon tax when they sell some of their products to some other countries?

Second, Van Rythoven himself acknowledges that "the US currently has no mechanism for carbon tariffs."

The White House under Democrat Joe Biden didn't impose carbon tariffs or a carbon tax. It's a safe bet President Donald Trump won't impose them either. And according to the federal government data, seventy-seven percent of our exports go to the United States.

Any country that wants to impose carbon tax tariffs would inevitably find itself in a trade war with the world's strongest economy, and a president (Trump) who has shown an inclination to punch back with tariffs of his own.

But even the US example downplays the extent to which other countries have opposed carbon taxes. In fact, about seventy percent of countries do not impose national carbon taxes, according to the World Bank, including large emitters like Russia, India and Brazil.

The threat of carbon tax tariffs from other countries doesn't warrant a Canadian government imposing carbon taxes on all Canadians. And certainly not pre-emptively. Plus, the vast majority of countries, including our largest trading partner, continue to reject carbon taxes.

The people who push for carbon taxes as a way to avoid potential tariffs are doing so for the simple fact they support carbon taxes and have lost the debate. It's a desperate tactic to try to convince Canadians they have no choice but to accept their favourite tax. And this reinforces the need for Canadians to remain vigilant even after the carbon tax is killed. The carbon taxers won't relent. They will keep trying to push carbon taxes down all of our throats.

The Death of the Carbon Tax

The carbon tax appears to be on death's doorstep. And that's because for more than a decade, ordinary Canadians from all walks of life pushed back against the elites in government, in political parties, in the universities, in the media and in big business.

The victory against Trudeau's carbon tax looks to be on the horizon. And there's one group of people who deserve credit for that victory. It's not any politician who seized on the momentum of opposition to the carbon tax. It's not even the CTF, which fought the carbon tax from day one. It's ordinary Canadians like you who never stopped pushing back against the elites and their carbon tax.

Thank you for fighting the carbon tax every step of the way.

As we move forward there are two important things to look out for.

The first is that academics, members of the media, politicians, government bureaucrats and big corporate interests in Canada and abroad will try to bring back the carbon tax. They're already calling for a "full strategic retreat" away from the consumer carbon tax to redouble their efforts to ensure the government keeps the hidden, industrial carbon tax or imposes some new form of carbon tax. The important thing for readers of this book to remember is that a carbon tax is a carbon tax is a carbon tax, no matter how the elites try to spin you.

The second is this: If Trudeau's carbon tax is sent to the dustbin, carbon tax activists will try to spin the narrative that carbon taxes are good, but the prime minister bungled the policy or didn't communicate it effectively.

Nothing could be further from the truth.

The carbon tax is not an economically efficient tax or environmental plan. There's nothing efficient about a tax that makes the necessities of life more expensive and does next to nothing for the global environment. This is especially true when the carbon tax is layered on top of a mountain of other energy taxes and regulations, like it was in Canada.

The carbon tax is not a market-based policy, as some academics try to sell conservative-leaning Canadians. There is nothing "market-based" about taxpayer-funded academics drawing fake models on their chalkboards, then advocating the government impose massive tax hikes on people because the real world doesn't reflect their imaginations.

The carbon tax and rebate scheme does not make life more affordable. Despite some pundits twisting themselves into pretzels trying to convince you otherwise, the entire point of the carbon tax is to make fuels like gasoline, diesel and natural gas more expensive. And the government can never take $20 from you and then magically give you a $50 back in return. And even if it could, even if the government did make you better off with carbon tax rebates, you wouldn't be getting that money from the government, you'd be taking that money from your neighbours or the small business down the street.

The carbon tax is not an environmental solution. A carbon tax that makes it more expensive for a Canadian mom to fuel up her minivan won't reduce emissions in China. A carbon tax that makes it more expensive for senior to stay warm in January won't reduce emissions in Russia. A carbon tax that makes it more expensive for a university student to afford food won't reduce emissions in the United States. It just leaves less money in Canadians' pockets to afford everything else.

The carbon tax is not a popular policy around the world. Our biggest trading partner, ally and competitor rejects carbon taxes, no matter which party is occupying the White House. And the vast majority of countries do not have national carbon taxes.

The carbon tax is bad not because Trudeau imposed it, nor because he bungled it. The carbon tax is bad because of what the carbon tax is. The carbon tax is a tax that does not work and makes all of our lives more expensive.

ABOUT THE CANADIAN TAXPAYERS FEDERATION

The Canadian Taxpayers Federation (CTF) fights for lower taxes, less waste and accountable government.

The CTF has been winning victories for taxpayers since its founding in 1990. The CTF has won referendums to stop tax hikes, exposed outrageously wasteful government spending, pushed for legislation to protect taxpayers and demanded balanced budgets. It has also been the loudest and most consistent critic fighting carbon taxes across Canada.

The CTF is a non-profit and non-partisan citizens' advocacy group headquartered in Regina, with offices in Ottawa and each region of the country. Every year, the CTF does thousands of media interviews, issues hundreds of news releases and columns, and, most importantly, connects directly with supporters through email, social media, live events and *The Taxpayer* magazine.

You can join the army of taxpayers fighting back against government waste and tax hikes by visiting the CTF's website: www.taxpayer.com